"Now, more than ever, we need to recognise and support just and sustainable community food initiatives. This book brings important issues of maintaining hope while staying with the trouble of enacting community food initiatives in a fair and just manner. It opens up our attention to matters of justice around food including as well as beyond procedural and distributional issues to essential matters of reparation."

Anna R. Davies, *FTCD, MRIA, Professor of Geography, Environment & Society, Director Environmental Governance Research Group, Department of Geography, School of Natural Sciences, Trinity College Dublin, the University of Dublin, Dublin 2, Ireland*

"How do the stories we tell about community food initiatives highlight or narrow their multiple ways of making culture and transforming political possibilities? This wide-ranging and comprehensively edited volume offers a variety of case studies that demonstrate the transformative work that community food initiatives envision and enact without shying away from acknowledging the ways that racial capitalism, hetero-patriarchy and neoliberalism constrain their approaches. This book reminds us that community food initiatives have much to offer as we combat the intersecting and inextricable social, environmental, and public health crises that shape this precarious moment."

Alison Hope Alkon, *Professor of Sociology, University of the Pacific*

Community Food Initiatives

This book examines a diverse range of community food initiatives in light of their everyday practices, innovations, and contestations.

While community food initiatives aim to tackle issues like food security, food waste, or food poverty, it is a cause for concern for many when they are framed as the next big "solution" to the problems of the current industrialised food system. They have been critiqued for being too neoliberal, elitist, and localist; for not challenging structural inequalities (e.g. racism, privilege, exclusion, colonialism, capitalism); and for reproducing these inequalities within their own contexts. This edited volume examines the everyday realities of community food initiatives, focusing on both their hopes and their troubles, their limitations and failures, but also their best intentions, missions, and models, alongside their capacity to create hope in difficult times. The stories presented in this book are grounded in contemporary theoretical debates on neoliberalism, diverse economies, food justice, community and inclusion, and social innovation, and help to sharpen these as conceptual tools for interrogating community food initiatives as sites of both hope and trouble. The novelty of this volume is its focus on the everyday doings of these initiatives in particular places and contexts, with different constraints and opportunities. This grounded, relational, and place-based approach allows us to move beyond more traditional framings in which community food initiatives are either applauded for their potential or criticised for their limitations. It enables researchers and practitioners to explore how community food initiatives can realise their potential for creating alternative food futures and generates innovative pathways for theorising the mutual interplay of food production and consumption.

This volume will be of great interest to students and scholars of critical food studies, food security, public health, and nutrition as well as human geographers, sociologists, and anthropologists with an interest in food.

Oona Morrow is an assistant professor of rural sociology at Wageningen University, the Netherlands.

Esther Veen is a professor of urban food issues at Aeres University of Applied Sciences Almere, the Netherlands.

Stefan Wahlen is a professor of food sociology at the University of Giessen, Germany.

Routledge Studies in Food, Society and the Environment

School Farms
Feeding and Educating Children
Edited by Alshimaa Aboelmakarem Farag, Samaa Badawi, Gurpinder Lalli and Maya Kamareddine

The Vegan Evolution
Transforming Diets and Agriculture
Gregory F. Tague

Food Loss and Waste Policy
From Theory to Practice
Edited by Simone Busetti and Noemi Pace

Rewilding Food and the Self
Critical Conversations from Europe
Edited by Tristan Fournier and Sébastien Dalgalarrondo

Critical Mapping for Sustainable Food Design
Food Security, Equity and Justice
Audrey G. Bennett and Jennifer A. Vokoun

Community Food Initiatives
A Critical Reparative Approach
Edited by Oona Morrow, Esther Veen, and Stefan Wahlen

Food Futures in Education and Society
Edited by Gurpinder Singh Lalli, Angela Turner, and Marion Rutland

The Soybean Through World History
Lessons for Sustainable Agrofood Systems
Matilda Baraibar Norberg and Lisa Deutsch

For more information about this series, please visit: www.routledge.com/books/series/RSFSE

Community Food Initiatives

A Critical Reparative Approach

Edited by
Oona Morrow, Esther Veen, and Stefan Wahlen

First published 2023
by Routledge
4 Park Square, Milton Park, Abingdon, Oxon OX14 4RN

and by Routledge
605 Third Avenue, New York, NY 10158

Routledge is an imprint of the Taylor & Francis Group, an informa business

British Library Cataloguing-in-Publication Data
A catalogue record for this book is available from the British Library

ISBN: 978-1-032-04902-1 (hbk)
ISBN: 978-1-032-04903-8 (pbk)
ISBN: 978-1-003-19508-5 (ebk)

DOI: 10.4324/9781003195085

Typeset in Goudy
by Apex CoVantage, LLC

Contents

Contributors

Stéphanie Eileen Domptail is a senior scientist and lecturer at the institute for agricultural policy and market research of the university in Gießen, where she investigates different facets of human-nature interaction in agricultural socio-ecological systems. She introduces concepts developed in ecological economics and political ecology to the analysis of agricultural and food systems. She is currently focussing on agroecology and the decolonisation of agricultural and development economics.

Jérémie Forney is professor for environmental anthropology at the Anthropology Institute, University of Neuchâtel. His research interests are centred on the governance of environmental issues related to food and agriculture. His approach is grounded in an ethnography of family farming and agri-food systems and focuses on the everyday perspectives and experiences of actors participating in complex governance assemblages. He engages more specifically with questions of autonomy, knowledge, and power around food production, circulation, transformation, and consumption. Part of his current scientific work is dedicated to the development of a theory of agri-environmental governance as an assemblage.

Marion Fresia is a professor at the Anthropology Institute, University of Neuchâtel. She has long-lasting research experience on forced migration, asylum bureaucracies, and humanitarian aid, on which she has extensively published. Since 2015, she has broadened her scientific interests to sustainable development and ecological transition initiatives, with case studies in West Africa and Switzerland. She has co-directed a research project on alternative agro-food networks in Switzerland as well as a research action project on the use of insects as an alternative source of proteins in smallholder farms in three west African countries. Currently, she is conducting extensive fieldwork on "ecovillage" initiatives in Senegal.

Oona Morrow is an assistant professor in food sociology in the Rural Sociology Group at Wageningen University. Her work is broadly concerned with the economic politics of everyday life, a theme she explores through the practice and politics of food provisioning in cities, communities, and households. Her writings on diverse economies, commoning, care, and food sharing have been published in *Transactions of the Institute of British Geographers*; *Gender, Place and Culture*; *Rethinking Marxism*; *Geoforum*; and *Urban Geography*.

Patrizia Pugliese is an agro-economist, a senior researcher, and a development officer at CIHEAM Bari, Italy. She holds a PhD in Policies for Sustainable Development and an MSc in Rural Resources and Countryside Management. Pugliese has 25 years of experience in research, training, and development cooperation work on rural development and sustainable and organic agriculture in the EU and developing countries. She has specific expertise in the Mediterranean region and focuses on rural economy diversification, sustainable and inclusive value chain development, and place-based and eco-regional approaches to territorial development. Her current fields of interest and work include rural women's empowerment and meaningful participation in agri-food value chains and the rural economy, agroecological transitions, and organic food systems. Pugliese authored various scientific contributions and participated in different EU-funded research projects. She is a member of the CIHEAM Bari Secretariat of the Mediterranean Organic Agriculture Network (MOAN) and a focal point for the CIHEAM Bari Institute of CIHEAM Corporate Working Group on Gender.

Marie Reine Bteich is a Project Officer – Field and Desk Researcher at the Mediterranean Agronomic Institute of Bari, Italy. She is an agronomist, holding an MSc in Mediterranean Organic Farming and a PhD in Engineering, Architecture, and Economics for the Urban and Rural Environmental Development. Bteich has 18 years of research expertise on organic farming, training, and international cooperation projects. She studies the socio-economic aspects of sustainable agriculture and rural development, trade, and supply chain within the EU and MENA countries. She is involved in several research and cooperation projects in the EU, MENA, and Balkan countries dealing with organic agriculture, sustainable rural development, and food security. Her focus is on gender-related issues, as well as sustainable value chain development in disadvantaged and conflict areas. She is also a member of the CIHEAM Bari secretariat of the Mediterranean Organic Agriculture Network (MOAN).

Marit Rosol is an urban and economic geographer and works as Chair of Economic Geography at the University of Wuerzburg, Germany. Previously, she held a Canada Research Chair and was Professor of Geography at the University of Calgary, Canada. She received her PhD from Humboldt-Universität zu Berlin and her Habilitation from Goethe-University Frankfurt, Germany. Her current research centres on the geographies of alternative food and alternative economic practices, as well as urban-based food movements. She has also published widely on housing, participation, urban governance, and urban gardening and contributed theoretically to debates on governmentality, political economy, (urban) political ecology, and hegemony.

Cosimo Rota is a data scientist consultant at the Mediterranean Agronomic Institute of Bari, Italy. He's an economist, holding a PhD in Agri-Food Politics and Economics and an MSc in Fair Trade and Organic Certification of the Agri-Food

Systems from Bologna University. As a researcher, Rota's field of interest relates to organic districts, food economics, supply chain management, rural development, and sustainability. He is involved in several international research projects as an associate project coordinator and data analyst. As a consultant, he works with the private and public sectors in support of project investigation, design, and development.

Felix Schilling holds a bachelor's degree in culture and technology from Brandenburg University of Technology Cottbus-Senftenberg, Germany. He completed his master's studies in agricultural and resource economics at Justus Liebig University Giessen, Germany, and agricultural sciences at Ghent University, Belgium. Felix works as a policy advisor and specialist for monitoring and evaluation in international development cooperation with a focus on land governance and rural development.

Anne Siebert is a postdoctoral researcher and lecturer at the Institute of Development Research and Development Policy (IEE), Ruhr-University Bochum, Germany. She obtained a joint PhD degree in International Development Studies from the IEE and the Institute of Social Studies, Erasmus University Rotterdam, the Netherlands. Her research experience and interest revolve around food politics and social movements, and how these have shaped dominant agri-food systems, governance, rural-urban interlinkages, as well as research methodology.

Elaine Swan is based at the Future of Work Hub at the University of Sussex Business School. Using feminism and critical race theories on food, she is interested in the intersections of race, gender, and cultural politics of food. She is currently co-PI on a UKRI project on transforming food systems for disadvantaged communities, working alongside community researchers and groups in Tower Hamlets in London, to value and improve food lives.

Lillian Tait is a PhD candidate at Macquarie University, working with Ngalakgan and Ritharrŋu co-researchers documenting the oral histories of Urapunga, Ngalakgan Country, Australia. She has a background in community arts and development studies. A paper based on Tait's master's thesis is published in the *Journal of Social and Cultural Geography*.

Hanne Torjusen works as a postdoctoral researcher at Consumption Research Norway (SIFO) at the Oslo Metropolitan University. She holds a PhD in Nutrition from the Faculty of Medicine at the University of Oslo. Her research interests include organic food and health, and sustainable consumption and production, particularly with a food systems perspective. Torjusen has participated in several EU-funded projects with interdisciplinary approaches. Current research projects are focused on the role of diversity in sustainable food systems and comparing and learning between the value chains for food and fibre/textiles with regard to sustainable stewardship of land and the practice of rangeland grazing.

Esther Veen works as a professor of urban food issues at Aeres University of Applied Sciences Almere, the Netherlands. Her focus is on food routines in the multicultural food environment of Almere. Veen holds an MSc in International Development Studies and a PhD from the Rural Sociology Group at Wageningen University in the Netherlands, based on research on community gardening, social cohesion, and alternative food networks. Veen has over ten years of research experience on urban agriculture, multifunctional agriculture, and healthy green cities, working in various positions at Wageningen University and Research. In her current position, she works closely together with Flevo Campus, an applied research institute with a focus on urban food issues in the Almere context.

Gunnar Vittersø works as senior researcher at Consumption Research Norway (SIFO) at the Oslo Metropolitan University. He holds a PhD in Human Geography from the University of Oslo. His research interests are in studies of sustainable consumption and production with a special focus on social transformation of the food system including studies of everyday food practices and alternative food networks. Vittersø has participated in several EU funded projects with interdisciplinary approaches among others including methods on involvement of citizens and stakeholders in research projects through deliberative processes.

Julien Vuilleumier is a social anthropologist and has been a doctoral student in the project "Alternative agro-food networks: innovative integration of sustainable eating habits and food production?" (NRP69, Swiss National Science Foundation). He investigated the three local contract farming initiatives of this "Food baskets project" through ethnographic fieldwork and participant observation. Developing connections between research, civil society participation, and political involvement, he is interested in social ecology, transition, and cultural participation. He is currently a scientific collaborator at the Swiss Federal Office of Culture, specialised in the fields of intangible cultural heritage, cultural diversity, and cultural routes.

Stefan Wahlen is professor of food sociology at the University of Giessen. After studying food and household studies at the University of Bonn, he completed his Ph.D. in consumer economics at the University of Helsinki (Finland). Following that, he worked at the chair group for Sociology of Consumption and Households at Wageningen University (the Netherlands). Current research focuses on food cultures and eating, as well as organisational and socio-political dimensions of food. Stefan is currently coordinating the European ERA-net project "FOOdIVERSE," which aims to uncover the role of diversity for a more sustainable and resilient food system. He is editor of the newly established journal *Consumption and Society* published by Bristol University Press.

Miriam Williams is a senior lecturer in geography and planning in the Macquarie School of Social Sciences at Macquarie University, Australia. As an urban geographer, her work focuses on everyday practices of care, justice, sustainability, diverse economies, and commons in the city. Her work has been published in international journals, including *Antipode*, *Urban Studies*, *Social and Cultural*

Geography, *Cities*, and *Area*. She is a member of the Community Economies Institute and is currently working on a project documenting community food provisioning initiatives in Sydney.

Fatima Zohra Sabrane is a PhD candidate at Seoul National University, South Korea, majoring in Agricultural and Resource Economics. She completed her Master of Science in Mediterranean Organic Agriculture from the Mediterranean Agronomic Institute of Bari, Italy. She also graduated as an Agricultural Engineer in Management of Plant Production and Environment from the Hassan II Institute of Agronomy and Veterinary Medicine in Morocco. Her Master of Science thesis on the Vulnerability Assessment of Urban and Peri-Urban Agriculture (UPA) organic projects in Morocco allowed her to open up on more socio-economic topics and led her to switch major from plant production to economics in her PhD programme. Currently, Fatima Zohra is working on rural households' income diversification issues and social network analysis studies. On a broader scale, she is also interested in environmental valuation methods, farmers' risk attitudes, impact evaluation, and development economics studies.

1 A critical reparative approach towards understanding community food initiatives

Acknowledging hopes and troubles

Oona Morrow, Esther Veen, and Stefan Wahlen

Introduction

The origins of this book are in a thematic session the editors organised at the RGS-IBG (Royal Geographical Society – Institute of British Geographers) Annual International Conference in London, in August 2019. The session was entitled "Cultivating hope while getting into trouble with Community Food Initiatives." Drawing inspiration from the main conference theme "from geographies of trouble to geographies of hope," we convened a session around community food initiatives (CFIs), asking presenters to discuss both the hopes and troubles of CFIs. Although hope and trouble were not specifically associated with the literature on community food initiatives, during the conference session and subsequent development of this edited volume, this conceptual pairing proved a particularly promising entry point for discussing and understanding CFIs. While the various presenters at the session interpreted the terms in different ways, the collective focus on both hope and trouble sparked inspiration and enthusiasm. The session sported lively debates and a general feeling of working towards a shared goal of better understanding CFIs in all their facets. Collectively, we were interested in the ambivalence that hope and trouble inspired and the perspective they offered when juxtaposed. In the years that followed, the unprecedented COVID-19 pandemic disrupted the global food system and everyday life, deepening existing social and economic inequalities. Communities in the majority and minority worlds experienced a sharp rise in food insecurity, food loss and waste, and loneliness, while everyday routines around food provisioning became incredibly complex. Community food initiatives rose to meet these challenges in creative ways through emergency food distribution, short food supply chains, and community meals and fridges, and found new ways to connect people and food. The flourishing of CFIs during this ongoing public health crisis gives us reason to be hopeful that a different kind of food system is possible.

The concepts of hope and trouble are guiding themes in this edited volume. We associate hope with optimism, a reparative stance, a politics of possibility, and an ethic of care and repair. This deliberately hopeful and asset-based approach to research and practice is exemplified by Gibson-Graham and the community economies institute whose work has shown how CFIs are in fact helping us to imagine and enact more-than-capitalist food economies (Gibson-Graham, 2006; Dixon,

DOI: 10.4324/9781003195085-1

2011; Sarmiento, 2017). A critical gaze, on the other hand, is more attuned to the insipient ways in which CFIs can also reproduce the very power structures and social inequalities they oppose (see, e.g. Slocum, 2007; Guthman, 2008). Trouble is often linked to a critique of the status quo and taken for granted. This can be expressed in the mischievous joy of "trouble making" as well as in cynicism and despair. Rather than choosing between a generous and reparative stance or a critical and paranoid gaze, we advocate for a more ambivalent approach to hope and trouble. This is best captured by thinkers like Lauren Berlant (2011), who cautions against "cruel optimism," in which hopeful desires can also sustain our suffering, and Donna Haraway (2016), whose work is a constant reminder of the impossibility of being purely oppositional or existing outside structures of power. Her invitation to "stay with the trouble" has created space for ambivalence and unknowing across the sciences. Translating this approach to CFIs means being sensitive to the numerous ways in which such initiatives are sites of contradiction that are interconnected with various aspects and challenges in the food system. We call this a critical-reparative approach: a way of knowing and understanding that highlights positive change, without shying away from describing the things that go wrong. The nuanced pictures that scholars paint in this book are helpful in advancing CFIs through a better understanding of their workings and developments and in generating the kinds of sustainability transformations that are so urgently needed. With this book and its somewhat eclectic collection of approaches towards hope and trouble, we aim to inspire the reader to develop a critical reparative approach towards understanding community food initiatives.

CFIs are incredibly diverse in terms of their mission and aims, organisational models, economies, politics, participants, geographies, ethics, and values. This diversity is reflected in the mosaic of social theories, concepts, and methodologies that are used to make sense of them. "Strong" theories often support the work of explaining and evaluating how alternative, sustainable, equitable and just, or transformative such initiatives are. This can lead to a selection bias that favours particular initiatives that support a dominant theory or are exemplary in terms of their innovation, success, or failure. By conceptualising CFIs as a dynamic field, rife with contradictions and tensions, we invited contributors to "get into trouble" with community food initiatives and contribute to the epistemic diversity of critical food studies, while unsettling the certainty found in strong critiques. In other words, we ask critical food scholars to stay a little longer with the trouble, rest in unknowing, and resist the pull of ironclad conclusions.

In our call for contributions, we asked authors to focus on what works for the specific CFI they studied while staying critical, to help readers understand difficulties without neglecting what can be celebrated. We began the process of editing this book and its individual chapters with a broad and flexible idea of what a critical reparative approach might entail, inviting authors to explore and theorise both hopes and troubles in community food initiatives. Ultimately, each author developed their own approach in response to our invitation, based on their interpretation of the themes of hope and trouble in their particular context. This led to the variety of theoretical entry points found in this book. For instance, Felix Schilling,

Stefan Wahlen and Stéphanie Eileen Domptail consider a moral economy perspective to describe how community food initiatives are (un-)making community. Patrizia Pugliese, Cosimo Rota, Fatima Zohra Sabrane, Marie Reine Bteich, and Esther Veen, on the other hand, use concepts of vulnerability and resilience to understand the hopes and trouble associated with CFIs. Gunnar Vittersø and Hanne Torjusen turn to convention theory to explain that people have different starting points for getting involved in a CFI, which leads to both difficulties and opportunities. And Jérémie Forney, Julien Vuilleumier, and Marion Fresia use the "promise of difference" to understand the hopes of local contract farming, and their relative success in realising different values and agricultural and economic practices.

The variety of approaches in this book contribute to the rich mosaic of theories and concepts found in critical food studies. The critical reparative approach that underpins this volume supports the use of a variety of theories and conceptual starting points that are fit for the purpose of enhancing our understanding of CFIs. A critical reparative approach is therefore a "weak" theory (Sedgwick, 2003; Gibson-Graham, 2006) that seeks to describe without explaining too much. In this vein, scholars approach CFIs with a broad and open mind to describe what is empirically happening. A critical reparative approach looks for, carves out, and brings about the small pearls of hope without belittling them. Nevertheless, the approach is also honest about the troubles CFIs face and have to deal with. As such, a critical reparative approach offers a mindset and family of concepts that aim to better understand the collective endeavours that are making change in the current industrialised food system.

The aim of this chapter is to introduce a **critical reparative approach** towards understanding CFIs. This necessitates navigating epistemological tensions and conceptual synergies in the field of critical food studies and consolidating multiple approaches to understanding CFIs. We do this by drawing on existing theoretical conceptualisations, empirical research, and approaches to challenges in the current food system that appear in the chapters of this volume. In the next section, we delineate what we understand as community food initiatives and why we think such an understanding is needed. Subsequently, we discuss how a critical reparative approach might take shape by using the lenses of hope and trouble. We explain how hope and trouble serve as unifying themes throughout the book. Such a critical reparative way of knowing allows a diverse set of approaches for answering old and new questions. This then leads to the third section, in which we outline how the individual chapters carry the threads of hope and trouble.

What are Community Food Initiatives?

Community Food Initiatives (CFIs) are driven by the specific needs, values, and concerns of people in different places and contexts who collectively come together to realise, rework, and challenge food systems. CFIs span rural and urban, consumption and production, alternative and mainstream, charity, mutual aid, and (social) entrepreneurship. They may work with, consist of, or cater to the young and the old, the rich and the poor, migrants and long-standing residents, and various groups in between. CFIs may involve stakeholders and practices at any stage of the food system – from growing, to (re)distributing, to preparing, cooking, and eating, to composting. However, not all food initiatives are *community* food initiatives: CFIs are food initiatives that are realised through collective actions addressing the place-specific needs and available resources of different communities.

Some community food initiatives can also be described as alternative, or civic, food networks. Alternative food networks (AFNs), for example, are food networks that define themselves as being different, notably *better* (at realising quality, sustainability, justice, equity, transparency, democratic governance, and so on) than the mainstream food system, offering solutions to the problems associated with that global industrial food system. AFNs often concern more direct relationships between producers and consumers, for example, through short food supply chains (Goodman et al., 2012). While the term is commonly used in literature, it has been criticised for defining initiatives in terms of what they are not, and for drawing exclusionary boundaries between the mainstream and the alternative, while still focusing on market relationships (Holloway et al., 2007; Rosol, 2020). The concept of civic food networks, which Renting et al. (2012) proposed as an alternative, is broader in the sense that it also includes initiatives that focus on civic actions beyond direct producer-consumer relationships. However, the concept of civic food networks mainly considers initiatives associated with farming and agriculture. We have adopted the term community food initiative (CFI) throughout this book because we wanted to include initiatives that are both alternative and mainstream, located in rural, peri-urban, and urban areas, and involved in meeting a variety of needs through agricultural production, food consumption, cultural inclusion, and civic engagement. CFIs open up a diversity of geographic scales (local, short supply chains), locations (urban, rural), practices (growing, preparing, eating), and concerns (health, sustainability) across the food system to collective actions that are responsive to the needs, resources, and abilities of specific communities. This inclusive conceptualisation of CFIs supports our aim of strengthening the epistemic diversity of critical food studies by recognising the collective endeavours of communities.

Community food initiatives are also sites for making community and negotiating difference – through formal and informal institutions, cultural norms, and normative frames. However, community should not be understood in positive terms only. Following feminist scholars (Young, 1989; Gibson-Graham, 2006; Nancy, 1991; Sedgwick, 2003), we understand community as an ongoing process of becoming, at loose ends with itself. We do not equate community with locality or territory, and we are wary of the politics of community that assumes sameness and obscures difference – especially when it comes to people and the various (even conflicting) values, beliefs, and motivations they hold. Rather than assuming that CFIs are composed of coherent, bounded, or pre-existing communities, we explore CFIs as sites for creating and contesting communities. Such contestation as well as creativity is enacted through practices of food production, provisioning, governance, commoning, consumption, and negotiation across difference. This is inspired by the approach that Gibson-Graham developed to understand the negotiation of economic difference and interdependence in community economies, which has proven to be relevant to understanding community food initiatives (Wilson, 2012; Loh & Agyeman, 2019; Cameron & Wright, 2014; Cameron et al., 2014). Centring the ethical doing of community in CFIs helps us to better understand how they collectively advance or resist social change.

A critical reparative approach: acknowledging hope and trouble in community food initiatives

In the literature on community food initiatives, we observe quite polarised ways of knowing CFIs. On the one hand there is a reparative stance oriented towards hope and possibility (Gibson-Graham, 2006), and on the other there is a critical gaze oriented towards trouble and failure (Guthman, 2008). Rather than framing hope and trouble as mutually exclusive, we take a more ambivalent approach that is guided by the practices and lived realities of community food initiatives, where both hope and trouble are present and overlap. A critical reparative approach therefore offers a more nuanced picture of community food initiatives. We are certainly not the first to advocate for such critical-reparative ways of knowing. We are indebted to scholars such as Donna Haraway (2016), who invites us to "make a critical and joyful fuss . . . to stay with the trouble" (p. 31), writing that "the only way I know to do that is in generative joy, terror, and collective thinking" (p. 31). We are also following Lauren Berlant (2011) whose concept of cruel optimism alerts us to the troubles in hope and the ways in which optimism can contribute to sustaining desires that can become obstacles to our own flourishing. In this way, we can see cultivating critical hope and getting into trouble as necessary elements in transformative research, for confronting the challenge of our food systems without slipping into despair and cynicism. We recognise the importance of affect and emotion to theorising and knowing, the joy in critique, the uncertainty of possibility, and the seduction of structures and systems that offer comprehensive explanations. This prompts us to reflect on which concepts, approaches, and possible futures attract our affective investments.

The reparative stance takes the perspective of real utopias, prefigurative politics, a politics of possibility, or an ethics of care (Jarosz, 2011; Blay-Palmer et al., 2016; Pettersson & Tillmar, 2022; Forno & Wahlen, 2022; Williams & Sharp, 2023), emphasising how social change can not only be thought of in theory but also how the world can be actively maintained, repaired, or transformed. Real utopias and prefigurative politics help us see how initiatives seek to enact utopias in the here and now in order to oppose states of dominance (Brower, 2013; Stock et al., 2015; Forno & Wahlen, 2022). The focus is on making alternative pathways more real by making them more visible and on strategically reading for difference rather than dominance so as to preserve a space for alternative futures (Gibson-Graham, 2006). Scholars taking this approach have examined topics such as community economies, solidarity economies, alterity, and autonomy (Forno & Graziano, 2014; Cameron & Wright, 2014; Rosol, 2020; Loh & Agyeman, 2019; Beacham, 2018). Arguably, this reparative stance offers a rosy picture, including the potential benefits of CFIs and the changes they can bring about in the food system as we know it. The critical gaze, often referred to as critical food studies, alerts us to the difficulties and unanticipated side effects of community food initiatives. CFIs have been critiqued for being too white, elitist, and localist and for reproducing neoliberalism and neoliberal subjectivities, the violence of charity, and the dominance of capitalist food economies in their own contexts (Guthman, 2008; Alkon & Guthman, 2017; Alkon & McCullen, 2010; Slocum, 2007; Dupuis & Goodman, 2005; Born & Purcell, 2006). They have also been condemned for the ways in which intersecting race, class, and gender inequalities are being mobilised and reproduced in these spaces (Born & Purcell, 2006; Guthman, 2008; Zitcer, 2015; Hayes-Conroy & Hayes-Conroy, 2010; Alkon & McCullen, 2010; McClintock, 2018; Argüelles et al., 2017; Galt et al., 2016; Weiler et al., 2016; Tregear, 2011; Lockie, 2013; Allen, 2008).

Empirically, studies with either a reparative or a critical approach show significant overlap when it comes to creating detailed case studies of AFNs, community gardens, urban farms, community food hubs, mutual aid organisations, food aid organisations, community supported agriculture, and food cooperatives. These studies differ in their focus and in what scholars notice and describe about CFIs by looking at their data in different ways, using different conceptual entry points, and applying different theories to tell different stories. Each approach therefore has its limitations and blind spots. Reparative ways of knowing community food initiatives, for example, infrequently attend to structural inequalities around intersecting forms of oppression and social difference. More critical ways of knowing community food initiatives tend to overemphasise the totality of neoliberal capitalism, focusing on imperfections and moments of failure rather than positive potentials and effects. Moreover, CFIs themselves are constantly struggling with internal tensions, like where to focus their limited resources, which struggles to fight, and how to get things done in the meantime. Such nuances and contradictions are too often left out of academic scholarship when it remains narrowly focused on either critiquing the difficulties and failures or celebrating the benefits and possibilities.

While there is merit to choosing a critical or reparative approach, as they enable us to see what community food initiatives can offer and what complexities ought to be taken into account, such studies risk painting a picture that is too black and white, either championing community food initiatives uncritically or failing to see possibilities beyond difficulties. We contend therefore that neither approach is sufficient to truly understand CFIs and how they could help change the food system. The assumed opposition between a critical gaze and a reparative stance can create tensions that affect our abilities to engage in research that both advances alternatives to the capitalist food system and challenges structural inequalities. We realise that the characterisations we sketched – the Pollyanna who sees only possibility and the paranoid cynic who sees nothing but neoliberalism – are straw people. Nevertheless, the frequent manifestation of these straw-people in food studies scholarship is worth reflecting on. In this book, we use the metaphors of hope and trouble to symbolise two dominant ways of knowing. We do not use "hope" and "trouble" as theoretical concepts but as entry points that help us reflect on the difference that CFIs can make on the one hand (the hopes), and the difficulties that they encounter on the other (the troubles). Hope and trouble are therefore invitations to critical and reparative ways of knowing.

We thank the contributors for taking up this invitation and for raising a number of important questions. How do community food initiatives create hope, while they are facing the trouble of an unequal and unjust food system? What troubles can CFIs run into, after a hopeful start? What are useful ways of dealing with troubles, and to what extent do these relate to particular hopes? Can we recognise elements in CFIs that function as hopes for some groups but as troubles for others? How can these troubles be turned into hopes? In answering these and other questions, each chapter demonstrates how a critical reparative approach can be assembled. For instance, making use of Erik Olin Wright's (2019) varieties of anti-capitalism, Anne Siebert sketches a diversity of strategies to manoeuvring transformative aims, potentials, and challenges to erode capitalism. Miriam Williams and Lillian Tait deploy care as a lens to emphasise hope in troublesome ethical negotiations and report on the role that CFIs take in mediating flows of care. Elaine Swan troubles the everydayness of coloniality that is stirred through the hopeful walking tours led by the women environmental network, and Marit Rosol asks for integrating environmental as well as social issues overcoming the duality of hope and trouble. These examples show how both hope and trouble are used to demonstrate the workings of different CFIs.

This edited volume is not alone in seeking to navigate through the ambivalence of hope and trouble. Conceptually, we take inspiration from feminist economic geographer J.K. Gibson-Graham's approach to enacting community economies and queer theorist Eve Sedgwick's weak theory (Sedgwick, 2003). These and similar thinkers have convincingly made clear that it is important to consider what is at stake in the stories we tell, in terms of subjectivity, knowledge production, and world making. Empirically, we are inspired by a growing body of critical food studies research that exemplifies a critical reparative way of knowing (see, e.g. Blay-Palmer et al., 2016; Rose, 2018; Mares, 2013; Drake, 2014; Sarmiento, 2017; Hayes-Conroy, 2008; Dixon, 2011; Wright, 2015; Moragues-Faus, 2017; Alkon & Guthman, 2017). In this body of research, we see a vigilant examination of power, justice, and inequality in CFIs that refuses the essentialism of neoliberalism and instead allows for the complexity and indeterminacy of embodiment, affect, and performativity. Within this frame, CFIs are incomplete assemblages, spaces for ambivalence, always open to contestation, and being performed differently. We aim to strengthen this perspective through the publication of this volume.

Outline of the book

The chapters presented in this volume thus use a diversity of theories and concepts. Therefore, we do not claim that the critical reparative approach we offer is comprehensive or complete. Rather, we contend that in their totality, the chapters start outlining possible toolkits that underscore the social dynamics of structure, agency, and transformation in community food initiatives. The chapters present cases that encourage modes of feeling, thinking, and knowing about community food initiatives that enable scholars to embrace "critical hope" while "staying with the trouble", as necessary elements for transformative research. By doing this in different ways, the book explores what such an approach could look like and what theoretical entry points could be promising. The different concepts and theories used in this book offer clues for operationalising a critical reparative way of knowing. The value of such an endeavour lies, therefore, in reminding us to incorporate a broad approach to food studies and to move beyond dichotomous views on community food initiatives that emphasise good and evil, drivers and barriers, pros and cons, environmental and social issues, or pitfalls and advantages. It presents a way of thinking, knowing, and feeling, that is open and attuned to the hopes and troubles of community food initiatives.

This book brings together a broad range of studies on CFIs. The chapters by Rosol and by Williams and Tait demonstrate the broad diversity of CFIs. Other chapters report on more specific organisational forms. CFIs may be legal entities, as in the local contract farming initiatives presented in the chapter by Forney, Vuilleumier, and Fresia and some of the organic farms described by Pugliese and colleagues. They may also be cooperatives, as reported on in the chapter by Schilling and colleagues, as well as in that by Torjusen and Vittersø. Or, they are loosely connected networks of individuals, supported by a formal non-profit or social enterprise, such as the cooking initiatives described in the chapter by Oona Morrow and the community gardening walking tours portrayed by Swan. Sometimes they are not easy to categorise, as evident in the work of Siebert, who writes about a community-led urban agriculture initiative. However, what the initiatives in this diversity of formalisation and institutionalisation have in common is that they are organised in, by, and for communities. The dynamics of community making and unmaking within the initiatives are of interest because they illustrate what people can collectively realise.

The chapters also respond to a variety of concerns in communities, including environmental concern, social justice, racism and colonialism, food insecurity, food waste, poverty, and inequality. They address these concerns through specific practices like urban agriculture, food redistribution, local contract farming, community supported agriculture, cooperative purchasing and growing, urban food governance, shared cooking and eating, organic farming, walking tours, education, and activism. The book captures a diverse (although certainly not exhaustive) variety of CFIs across the globe in Europe, North America, Africa, and Australia. This variety in case studies (see Table 1.1) is brought together to understand both the hope and the trouble that these initiatives celebrate and struggle with.

Overview of concepts

While the chapters of this book report on very different community food initiatives, using different conceptual and theoretical approaches, they have a critical reparative viewpoint in common. The collection of examples highlights various issues that we cluster around three broader avenues of investigation, all demonstrating the need for different ways of understanding CFIs:

The first part of the book is concerned with social justice and inequality in urban contexts. It draws on a variety of concepts that assist in advancing a critical reparative approach.

Table 1.1 Overview of concepts

	Chapter	*Concept*	*Empirical case*	*Contribution to a critical reparative approach*
Part 1: CFIs addressing social injustices and inequalities in urban food				
2	Williams and Tait	Care thinking: relationality and connection and more-than-human entanglements	Food pantry/CSA	Caring for, about and with human as well as nonhuman others to create another food system
3	Pugliese et al.	Vulnerability/ resilience	Urban agriculture	Perturbation (risk/ vulnerability literature)
4	Siebert	Varieties of anti-capitalism	Urban agriculture initiative	Moving beyond isolated manoeuvres with the troubled agri-food system
5	Rosol	Environmental and social justice	Urban food, food policy councils	Combining social and environmental effort
Part 2: Cooperatives, cooperation, and concerns in CFIs				
6	Fourney et al.	Promise of difference	Local contract farming	
7	Torjusen and Vittersø	Conventions theory	Food cooperative	Difference in community, shared conventions
8	Schilling et al.	Moral economy	Community-supported agriculture	Community (un-) making
Part 3: Commensality, social gatherings, and food knowledge in CFIs				
9	Swan	Everyday environmentalism, colonialism	Women environmental network walking tours	
10	Morrow	Conviviality, eating (with) the other	Initiatives that promote cooking and eating together as a mode of cultural exchange	Focus on emotions of desire and discomfort in eating the other vs. eating with the other

Miriam Williams and Lillian Tait reflect upon the notion of care and how this shapes hopes and troubles. The chapter entitled "Caring in unequal worlds: tracing the hopes and troubles of Community Food Initiatives in Sydney" attends to two domains of care-thinking: first, relationality and connection and, second, more-than-human entanglements. From these two perspectives, Williams and Tait trace how care scholarship offers a unique lens for thinking about how communities are constituted and the complexities of caring in unequal worlds. Empirically, the chapter draws upon data from a research project that documented community food initiatives across metropolitan Sydney, Australia, in 2019. In particular, they analyse two initiatives: the Addison Road Community

Centre Organisation (ARCCO) Food Pantry and Five Serves Produce Community Support Agriculture Scheme (CSA). These two community food initiatives assist in reflecting hopes and troubles through a care lens. Paying attention to care as a vital practice constituting communities of CFIs reveals the expansive communities involving both human and non-human others that are cared for, cared about and care with CFIs.

The next chapter by **Patrizia Pugliese, Cosimo Rota, Fatima Zohra Sabrane, Marie Reine Bteich,** and **Esther Veen** is interested in the vulnerability and resilience of urban food initiatives in Morocco. Vulnerability of urban agriculture has not been described in this context before. In urban agriculture, perturbations such as stress can be considered as troubles, which assist in identifying challenges. People have agency and negotiate how they respond to troubles, depending on their situations. Some have more going for them in solving the problem. This approach helps the authors get out of the structure agency dichotomy, by looking at hope and trouble and at what is going on in the respective initiatives. The concept of vulnerability is useful for showing the structural, social, and environmental struggles people face in CFIs but the parallel concept of resilience helps understanding that people are differently positioned to face these struggles.

Anne Siebert's chapter was inspired by a community-led urban agriculture initiative in Cape Town (South Africa), trying to understand how alternative food providers can truly create hope through resisting the status quo and confronting related troubles. This reflects an everyday situation in which food initiatives view the dominant industrialised agri-food system critically but operate in niches and struggle to create truly revolutionary alternatives. The case study presented illuminates this interplay of hope as a way of proposing alternatives and its constant companion, trouble, enshrined in the country's agri-food system and highly unequal society. Siebert's critical reparative approach is inspired by Erik Olin Wright's (2019) varieties of anti-capitalism. She aims to analyse diverse strategies for challenging the dominant food system and related inequalities. For instance, the initiative's motivation to be self-sufficient overlaps with the strategy of escaping capitalism. Wright's considerations help to sketch out new perspectives and broaden the horizon beyond isolated manoeuvres within the troubled agri-food system.

Marit Rosol focuses in the subsequent chapter on a specific trouble of community food initiatives: their often limited focus on environmental sustainability, which is only one aspect of our current food systems' challenges. Hence, despite long-standing calls for integrated social and ecological transformations of our agri-food systems, in practice, community food initiatives often focus on either the environmental aspect – often in cooperation with producers – or the social justice aspect, often with a focus on food insecure (urban) consumers. Based on an analysis of the literature and empirical research in Germany and Canada, the chapter shows this prevalence of single-issue approaches and contextualises these findings through a conceptual literature review. Rosol then discusses two possible ways to overcome this problem: the Canadian concept of community food centres (CFCs), and food policy councils as a tool to envision and enact integrative food policies on a municipal level. While not shying away from critique and pointing out shortcomings and difficulties, these two examples are real attempts to overcome divisions and thus points of inspiration and hope.

The second part of the book addresses cooperation, conflict, and concerns in community food initiatives. All chapters in this section studied variations of food cooperatives that bring one or more producers and consumers together. Organisational forms of such cooperatives are more traditional cooperative contract farming, as well as community supported agriculture. The chapters emphasise the difficulties that such cooperatives struggle with and the successes they celebrate, using different theoretical starting points.

Jérémie Forney, Julien Vuilleumier, and Marion Fresia studied three local contract farming initiatives in Switzerland, starting from a "promise of difference" (Le Velly, 2017). The authors argue that it is important to first recognise the intentions of local contract farming without assuming that their operating methods should be radically different from mainstream food systems, and to only then examine the concrete changes brought about. Their empirical work draws on two elements that constitute essential vehicles for the rearrangements they observed: the contract that translates and formalises the promise of long-term commitment from consumers and the quality of the food. By means of these empirical investigations, the authors move beyond the debate about whether these networks are truly alternative or not, understanding them as particular collections of human and non-human actors that "open up spaces in which to enact a politics of possibility" (Harris, 2009, p. 58).

Hanne Torjusen and Gunnar Vittersø use convention theory to study a consumer cooperative in Norway. They advance two lines of inquiry to investigate how a consumer cooperative may contribute to food system sustainability. First, they analyse the importance of shared conventions related to environmental, economic, and social sustainability as a common basis for the *outcomes* of the activities within the CFI. The second line of inquiry is related to the *process*, by investigating the functioning and development of Vestfold Kooperativ. The chapter draws on conventions theory to analyse how shared conventions (hope) and contested issues (trouble) are perceived within the cooperative and how troubles eventually might be overcome. The chapter concludes that, in spite of the troubles that Vestfold Kooperativ faced, there are important lessons to be learnt from their experiences.

Felix Schilling, Stefan Wahlen, and Stéphanie Eileen Domptail use a moral economy perspective to understand why farmers participate in three community supported agriculture initiatives in Germany. Focusing on the role and position of farmers, they ask how the relevant roles and responsibilities in these initiatives are negotiated and distributed among farmers and consumers, and how they succeed in enacting farmers' objectives. The chapter reveals the hopes and troubles of CSA farmers, emphasising how, and to what extent, cooperation with consumers enables them to achieve their own agendas in the face of their dissatisfaction with the current conventional food system and its institutions. In addition, the authors investigate how the redistribution of responsibilities between producers and consumers contributes to job satisfaction and to the creation of new roles for CSA members.

The third part of the book underscores the role of commensality, social gatherings, and knowledge in community food initiatives in urban areas. The authors of the chapters highlight the importance of intersectional feminist, anti-racist, and decolonial perspectives for unravelling power dynamics in community food initiatives. Where the previous section focused on communities of farmers and consumers, this section focuses on urban communities around centred eating and sharing food.

Elaine Swan's chapter is focused on everyday environmentalism in decolonising food initiatives. Describing two community gardening walking tours, she brings attention to their unseen colonial histories, highlighting how colonialism lives in our everyday landscape. In everyday situations, racial histories and dynamics are easily ignored while they are, in reality, ever present. These neighbourhoods experience environmental racism: people have less access to green space and good food, and their knowledge is valued less. The author explains that while such coloniality and racism are cases of trouble, the walking tours themselves can be seen as hope and a source of inspiration. Hence, while the walking tours do not solve the trouble, they do enable a better understanding of the ecohabitus of white environmentalism and to sensitise people to that.

Oona Morrow examines two community cooking initiatives that seek to valorise the culinary heritage and foodways of migrants in New York City and Berlin, examining how cooks and eaters within these initiatives navigate the power asymmetries of their transactions in their shared desire for transformative culinary encounters. Both initiatives described use food knowledge, cooking, and eating to address the troubles of xenophobia. They offer migrants and locals the hope of cross-cultural friendships and social networks, as well as employment opportunities for people who are marginalised from the labour market and devalued in the food service industry. Morrow shows how the relative success of these initiatives depends on the emotional and performative labour of cooks who stage authenticity and identity, and the desires of an affluent group of consumers willing to pay a premium for these experiences.

Each of the chapters realises a critical reparative approach using different conceptual entry points, including the promise of difference, vulnerability and resilience, conventions theory, care, moral economy, everyday environmentalism, eating (with) the other, or the varieties of anti-capitalism. As the authors demonstrate in their respective chapters, each of these concepts can support a critical reparative approach to understanding CFIs. Nevertheless, there is still a great deal of diversity within this approach, with chapters painting CFIs in different gradations of hopefulness and critique.

Conclusion

By demonstrating the various ways in which researchers can look beyond a black and white picture – by focusing on shades of grey, pink, green, or other and by highlighting both hope and trouble – this book contributes to sharpening the theoretical and methodological tools of critical food studies. The collection of case studies demonstrates how a critical reparative approach enables a vision of CFIs that situates them in the real world, where troubles are not always easily overcome and dominant structures are indeed not always overthrown and sometimes even perpetuated, and where small victories are creating other possible worlds that can be detected, learned from, and brought to the fore. With this collection of chapters, we make a first start. In our reading of the chapters, a critical reparative approach centres on social inequality and social injustices that many, yet not all, community food initiatives address. The capacities of CFIs to progress on these issues are closely linked to their own context and community composition, as well as the historic patterns, current trends, and views on the future that frame the realm of possibility.

We invite future researchers to further define and refine the critical reparative approach(es) we have assembled here. An important question that remains is what a critical reparative approach means for how researchers and activists approach research and action: the questions to be asked, the methods to be taken, the way in which to analyse data, etc. Also, researchers can reflect on which theoretical starting points are most compatible with a critical reparative approach and what methodologies can be derived from the different conceptual approaches taken. Future research could also try and include different territories and places in the world. Although we tried to include case studies from all over the world in this book, most chapters focus on the minority world, with CFIs in the majority world understudied. Finally, taking a critical reparative approach seriously also means that researchers should try and learn not only from best practices but also from failures, as these often teach us not only about struggles but also about initial, remaining, and newly defined hopes.

References

Alkon, A. H., & Guthman, J. (Eds.). (2017). *The new food activism: Opposition, cooperation, collective action*. University of California Press.

Alkon, A. H., & McCullen, C. G. (2010). Whiteness and farmers markets: Performances, perpetuations . . . contestations? *Antipode*, *43*(4), 937–959. https://doi.org/10.1111/j.1467-8330.2010.00818.x

Allen, P. (2008). Mining for justice in the food system: Perceptions, practices, and possibilities. *Agriculture and Human Values*, 157–161. https://doi.org/10.1007/s10460-008-9120-6

Argüelles, L., Anguelovski, I., & Dinnie, E. (2017, August). Power and privilege in alternative civic practices: Examining imaginaries of change and embedded rationalities in community economies. *Geoforum*, 86, 30–41. https://doi.org/10.1016/j.geoforum.2017.08.013

Beacham, J. (2018). Organising food differently: Towards a more-than-human ethics of care for the anthropocene. *Organization*, *25*(4), 533–549. https://doi.org/10.1177/1350508418777893

Berlant, L. (2011). *Cruel optimism*. Duke University Press.

Blay-Palmer, A., Sonnino, R., & Custot, J. (2016). A food politics of the possible? Growing sustainable food systems through networks of knowledge. *Agriculture and Human Values*, *33*(1), 27–43. https://doi.org/10.1007/s10460-015-9592-0

Born, B., & Purcell, M. (2006). Avoiding the local trap: Scale and food systems in planning research. *Journal of Planning Education and Research*, *26*(2), 195–207. https://doi.org/10.1177/0739456X06291389

Brower, A. (2013). Agri-food activism and the imagination of the possible. *New Zealand Sociology*, *28*(4), 80–100.

Cameron, J., Gibson, K., & Hill, A. (2014). Cultivating hybrid collectives: Research methods for enacting community food economies in Australia and the Philippines. *Local Environment*, *19*(1), 118–132. http://doi.org/10.1080/13549839.2013.855892

Cameron, J., & Wright, S. (2014). Researching diverse food initiatives: From backyard and community gardens to international markets. *Local Environment*, *19*(1), 1–9. https://doi.org/10.1080/13549839.2013.835096

Dixon, J. (2011). Diverse food economies, multivariant capitalism, and the community dynamic shaping contemporary food systems. *Community Development Journal*, *46*(Suppl. 1), 20–35. https://doi.org/10.1093/cdj/bsq046

Drake, L. (2014). Governmentality in urban food production? Following "community" from intentions to outcomes. *Urban Geography*, *35*(2), 177–196. https://doi.org/10.1080/02723638.2013.871812

DuPuis, E. M., & Goodman, D. (2005). Should we go "home" to eat?: Toward a reflexive politics of localism. *Journal of Rural Studies*, *21*(3), 359–371. https://doi.org/10.1016/j.jrurstud.2005.05.011

Forno, F., & Graziano, P. (2014). Sustainable community movement organisations. *Journal of Consumer Culture*. https://doi.org/10.1177/1469540514526225

Forno, F., & Wahlen, S. (2022). Environmental activism and everyday life. In M. Grasso & M. Giugni (Eds.), *The Routledge handbook of environmental movements*. Routledge.

Galt, R. E., Bradley, K., Christensen, L., Van Soelen Kim, J., & Lobo, R. (2016). Eroding the community in community supported agriculture (CSA): Competition's effects in alternative food networks in California. *Sociologia Ruralis*, *56*(4), 491–512. https://doi.org/10.1111/soru.12102

Gibson-Graham, J. K. (2006). *A postcapitalist politics*. University of Minnesota Press.

Goodman, D., DuPuis, E. M., & Goodman, M. K. (2012). *Alternative food networks: Knowledge, practice, and politics* (1st ed.). Routledge. https://doi.org/10.4324/9780203804520

Guthman, J. (2008). Bringing good food to others: Investigating the subjects of alternative food practice. *Cultural Geographies*, *15*(4), 431–447. https://doi.org/10.1177/1474474008094315

Haraway, D. J. (2016). *Staying with the trouble*. Duke University Press.

Harris, E. (2009). Neoliberal subjectivities or a politics of the possible? Reading for difference in alternative food networks. *Area*, *41*(1), 55–63.

Hayes-Conroy, A., & Hayes-Conroy, J. (2010). Visceral difference: Variations in feeling (slow) food. *Environment and Planning A*, *42*(12), 2956–2971. https://doi.org/10.1068/a4365

Hayes-Conroy, J. (2008). Hope for community? Anarchism, exclusion, and the non-human realm. *Political Geography*, *27*(1), 29–34. https://doi.org/10.1016/j.polgeo.2007.09.001

Holloway, L., Kneafsey, M., Venn, L., Cox, R., Dowler, E., & Tuomainen, H. (2007). Possible food economies: A methodological framework for exploring food production – consumption relationships. *Sociologia Ruralis*, *19*(2), 6–15. https://doi.org/10.1111/j.1467-9523.2007.00427.x

Jarosz, L. (2011). Nourishing women: Toward a feminist political ecology of community supported agriculture in the United States. *Gender, Place, and Culture*, *18*(3), 307–326. https://doi.org/10.1080/0966369X.2011.565871

Le Velly, R. (2017). *Sociologie des systèmes alimentaires alternatifs: Une promesse de différence* (p. 200). Presses des Mines.

Lockie, S. (2013). Bastions of White privilege? Reflections on the racialization of alternative food networks. *International Journal of Sociology of Agriculture & Food*, *23*(3), 409–418.

Loh, P., & Agyeman, J. (2019). Urban food sharing and the emerging Boston food solidarity economy. *Geoforum*, 99, 213–222. https://doi.org/10.1016/j.geoforum.2018.08.017

Mares, T. M. (2013). "Here we have the food bank": Latino/a immigration and the contradictions of emergency food. *Food and Foodways*, *21*(1), 1–21. https://doi.org/10.1080/07409710.2013.764783

McClintock, N. (2018). Cultivating (a) sustainability capital: Urban agriculture, ecogentrification, and the uneven valorization of social reproduction. *Annals of the American Association of Geographers*. https://doi.org/10.1080/24694452.2017.1365582

Moragues-Faus, A. (2017). Emancipatory or neoliberal food politics? Exploring the "politics of collectivity" of buying groups in the search for egalitarian food democracies. *Antipode*, 49(2), 455–476. https://doi.org/10.1111/anti.12274

Nancy, J.-L. (1991). *The inoperative community*. The University of Minnesota Press.

Pettersson, K., & Tillmar, M. (2022). Working from the heart – cultivating feminist care ethics through care farming in Sweden. *Gender, Place & Culture*, 1–21.

Renting, H., Schermer, M., & Rossi, A. (2012). Building food democracy: Exploring civic food networks and newly emerging forms of food citizenship. *The International Journal of Sociology of Agriculture and Food*, *19*(3), 289–307. https://doi.org/10.48416/ijsaf.v19i3.206

Rose, M. (2018). The diversity we are given: Community economies and the promise of Bataille. Antipode, 1–18. https://doi.org/10.1111/anti.12424

Rosol, M. (2020). On the significance of alternative economic practices: Reconceptualizing alterity in alternative food networks. *Economic Geography*, 96(1), 52–76. https://doi.org/10.1080/00130095.2019.1701430

Sarmiento, E. R. (2017). Synergies in alternative food network research: Embodiment, diverse economies, and more-than-human food geographies. *Agriculture and Human Values*, *34*(2), 485–497. https://doi.org/10.1007/s10460-016-9753-9

Sedgwick, E. K. (2003). *Touching feeling: Affect, pedagogy, performativity*. Duke University Press.

Slocum, R. (2007). Whiteness, space and alternative food practice. *Geoforum*, *38*(3), 520–533. https://doi.org/10.1016/j.geoforum.2006.10.006

Stock, P., Carolan, M., & Rosin, C. (2015). *Food utopias: Reimagining citizenship, ethics and community*. Routledge.

Tregear, A. (2011). Progressing knowledge in alternative and local food networks: Critical reflections and a research agenda. *Journal of Rural Studies*, *27*(4), 419–430. https://doi.org/10.1016/j.jrurstud.2011.06.003

Weiler, A. M., Otero, G., & Wittman, H. (2016). Rock stars and bad apples: Moral economies of alternative food networks and precarious farm work regimes. *Antipode*, *48*(4), 1140–1162. https://doi.org/10.1111/anti.12221

Williams, M. J., & Sharp, E. L. (2023). Feminist ethics of care in urban food governance. In *Routledge handbook of urban food governance* (pp. 78–91). Routledge.

Wilson, A. D. (2012). Beyond alternative: Exploring the potential for autonomous food spaces. *Antipode*, *45*(3), 719–737. https://doi.org/10.1111/j.1467-8330.2012.01020.x

Wright, E. O. (2019). *How to be an anticapitalist in the twenty-first century* (1st ed.). Verso.

Wright, S. (2015). More-than-human, emergent belongings: A weak theory approach. *Progress in Human Geography*, *39*(4), 391–411. https://doi.org/10.1177/0309132514537132

Young, I. M. (1989). The ideal of community and the politics of difference. In Linda J. Nicholson (Ed.), *Feminism/Postmodernism* (pp. 300–323). Routledge.

Zitcer, A. (2015). Food co-ops and the paradox of exclusivity. *Antipode*, *47*(3), 812–828. https://doi.org/10.1111/anti.12129

Part 1

CFIs addressing social injustices and inequalities in urban food

2 Caring in unequal worlds

Tracing the hopes and troubles of Community Food Initiatives in Sydney

Miriam Williams and Lillian Tait

Introduction

The aim of this chapter is to reflect on how paying attention to care might help us understand the hopes and troubles that shape Community Food Initiatives (CFIs). Care is understood as a "species activity that includes everything that we do to maintain, continue, and repair our 'world' so that we can live in it as well as possible" (Fisher and Tronto cited by Tronto, 1993, p. 113). The connections between practices of care and the provisioning of food have long piqued the interests of human geographers and critical food scholars. Resulting scholarship has been concerned with the ways care may operate through alternative food networks (Cox, 2010; Goodman et al., 2010; Kneafsey et al., 2008); urban food commons (Morrow, 2019; Morrow & Martin, 2019); spaces of food relief (Bedore, 2018; Cloke et al., 2017; Conradson, 2003; Johnsen et al., 2005; Williams, 2017a); "care-full food systems transformations" (Sharp, 2019); and practices of eating (Abbots et al., 2015); amongst many others. Care involves five phases: caring for, caring about, care giving, care receiving, and caring with (Tronto, 1993, 2013). Care is not simple or easy; it can be hard work, messy, relentless, open-ended, tiring, and troubling (Puig de la Bellacasa, 2017). Focusing on the presence or absence of care reveals needs, wants, privileges, inequalities, exclusions, and inclusions. Those who do the work of caring in societies are often marginalised and bear the "material consequences" of care (Puig de la Bellacasa, 2010, p. 165). A feminist ethics of care has played a role as an evaluative framing for scholars seeking to understand how worlds might be transformed to become more caring (Power & Williams, 2019; Williams, 2020). Such work is attentive to the uneven capacities, power relations, and ways in which responsibilities to and for care are enacted at the same time as emphasising the possibility of an ethics of care to refashion worlds so that they might become more caring (Beacham, 2018; Williams, 2020).

Our work is informed by scholarship that attends to the inequalities and injustices that shape people's caring capacities, who takes responsibility to/for care whilst also emphasising the potential of care to bring "as well as possible" worlds into being through maintaining, continuing, repairing and transforming unequal worlds (Power, 2019; Power & Williams, 2019; Williams, 2020). We explore how such an approach, which centres the role of caring as relational practices that concerns

DOI: 10.4324/9781003195085-3

more-than-human others, may be useful for scholars interested in understanding CFIs through a critical-reparative approach that shapes this book. The critical-reparative approach woven throughout this book involves being attentive to both the hopes and troubles of CFIs. Within the context of this chapter, we argue that a focus on care enables us to apply the critical-reparative approach as we join other scholars who performatively and politically highlight the importance of care in our worlds as a way to hold troubles and hope in tension (Power & Williams, 2019; Williams, 2020). Throughout, we emphasise the context of inequality and unevenness that pervades the work of CFIs and shapes their caring capacities in unequal worlds (Power, 2019): worlds of abundance and scarcity.

The chapter is structured as follows. First, it attends to two domains of care-thinking – (1) relationality and connection and (2) more-than-human entanglements – to trace the insights they offer about how communities are constituted and the complexities of caring in unequal worlds. Second, we discuss the research context and methods, which involved drawing upon data from a research project that documented CFIs across metropolitan Sydney. Third, we offer an in-depth analysis of two initiatives: the Addison Road Community Centre Organisation (ARCCO) Food Pantry and Five Serves Produce Community Support Agriculture Scheme (CSA). We emphasise the ways in which care is practiced through relational connections and more-than-human others to reveal the complexities of hope and trouble evident in the function of CFIs in an unequal world.

Relational more-than-human connections

In this section, we engage with two key domains of care-thinking to trace the insights they offer into the relational connections that constitute CFIs and the hopes and troubles that emerge. First, we attend to the relationality and connection that facilitate the actors that constitute CFIs to care with and through CFIs, discussing how a focus on care assists us in understanding the expansive communities that are cared for through and by CFIs and the limits of such care. Second, we attend to how more-than-human others are made visible through care-thinking as actors that are integral to care through CFIs, which highlight caring capacities in unequal worlds.

Relationality and connections

> For not only do relations involve care, care is itself relational.
>
> (Puig de la Bellacasa, 2012, p. 199)

As a practice and ethics, care is relational (Puig de la Bellacasa, 2012, p. 199). Care creates relationships and is "an affective force" (Puig de la Bellacasa, 2010, p. 164). Relationality and connections are central to our understandings of practices of care in three ways. First, care-thinking creates an expansive understanding of community due to its foundation in a conceptualisation of self as relationally connected and inter-dependent upon others for survival (Beacham, 2018; Lawson, 2007). Acknowledging our commonality of being enables us to "recognise the interdependency joining us all together" (Beacham, 2018, p. 543). Such ways of relating to the world have been practiced by some First Nations communities around the world for millennia, emerging from an ontological perspective that understands all beings as inherently connected in bounded relationships of mutual dependency and responsibility (Bawaka Country et al., 2016; Wilson, 2008). Relationality is a practice shaping understandings of the world and others, as well as a responsibility to be accountable to these more-than-human relations through our actions and our ways of caring and relating (Bawaka Country et al., 2016; Tynan, 2021; Wilson, 2008). Paying attention to care therefore allows us to trace these expansive connections of responsibility that might be practiced in CFIs.

Second, tracing practices of care makes visible who becomes part of the community that cares for through and is cared for by a CFI. Like Nancy (1991), we understand community as "always-in-process" (Popke, 2003, p. 310) and CFIs as assisting people to negotiate the complexity of what it means to always be responsible to others due to the interconnections made known by food (Beacham, 2018, p. 543). CFIs operate to make others increasingly visible by creating more direct or tangible linkages between producers and consumers in order to further care for multiple actors across food systems – as care givers or receivers (Beacham, 2018; Kneafsey et al., 2008; McEwan & Goodman, 2010). For example, "alternative" or diverse food schemes such as fair trade have sought to reconnect or create more transparent connections or awareness of the needs of producers (Beacham, 2018; Kneafsey et al., 2008). Improving the transparency and connections between producers and consumers, as a way to "know" producers is seen as a way to provoke care for others. In their work on food system alternatives, Dowler et al. (2010) found three care-full motivations for participation in CFIs, which included ensuring the "transparency and integrity in food systems, including matters of science and governance" (Dowler et al., 2010, p. 213).

At times, the people who care for each other through CFIs are strangers, unknown to each other due to the mediated nature of care-giving enabled (Alam & Houston, 2020). For example, in their study of care as alternate infrastructure, Alam and Houston (2020) explain how a roadside fruit and vegetable shelf enables people to remain anonymous. The anonymity fostered here enables care to be provided "without disclosing the identity and vulnerability of either party" (Alam & Houston, 2020, p. 4). Knowing who provides or is receiving the care provisioned through the roadside vegetable shelf is not necessary, as "there is a sense of recognition of everybody's ability to contribute through sharing what they have 'spare'" (Alam & Houston, 2020, p. 4). What becomes evident in this example, however, is the way in which the free fruit and vegetable shelf acts as a mediator of care that enables people who are able to care for and about each other.

Third, tracing care also makes visible the exclusions and limitations placed on who is able to be cared for by and through CFIs. For example, food relief initiatives are often dependent on the redistribution of food waste from retailers and have limited capacity to consider the sustainability and ethical considerations of food supply (Williams & Tait, 2022). In contrast, food cooperatives and ethical food box schemes are often guided by particular buying principles that ensure the food supplied cares for farmers and the environment (Williams, 2017b; Williams & Tait, 2022), but can have difficulty in making such food affordable for people on low incomes. People may also be excluded from participating in CFIs for various reasons, including their socio-economic status, which restricts their ability to afford ethically produced food. Healthy and sustainably produced food is often less accessible to low-income communities, which excludes them from participating in alternative food networks and initiatives with a focus on "organic" food (Alkon & Guthman, 2017, pp. 7–8). Subscription-based systems such as CSAs enable those who can afford the upfront cost to purchase organic food and practice care (Kneafsey et al., 2008). CFIs vary in their capacity to care for more-than-human others, despite these others being vital to the care practiced through CFIs.

More-than-human care

Increasingly, scholars have become attuned to the more-than-human relations of care and the ways in which more-than human others are always present in caring relations (Alam & Houston, 2020). Communities that constitute CFIs are dynamic and comprise both more-than-humans and human others (Beacham, 2018; Gibson-Graham, 2006). Care is more than a human concern, with more-than-human others knowingly and unknowingly reciprocating care (Puig de la Bellacasa, 2017). Attending to care makes more-than-human others visible and reinforces our collective interdependence in two ways. First, it makes more-than-human others visible as both objects and subjects of care (Power & Williams, 2019). Care is a vital practice for the continuation of life and provides a means through which to enact responsibility towards more-than-human others (Beacham, 2018; Kneafsey et al., 2008). CFIs may be attuned to care for soils, worms, microbes, bees, plants, rivers, and waterways involved in the collective that are cared for with and through the CFI. For example, Beacham (2018) discusses how soil contains nutrients that nourish plants, which humans are dependent on for survival. Not all CFIs care for or have direct relationships with soil, for example, initiatives that redistribute food waste to people experiencing food insecurity. Yet, people practice care for more-than-human others and are cared for by more-than-human others through their involvement in CFIs such as the CSAs studied by Beacham (2018).

Second, "object, bodies, buildings, or materials" (Power & Williams, 2019, p. 3) might enable care, enhance caring capacity, or care alongside humans. Care might be woven through particular spaces of care or caring such as the home as a site of domestic provisioning (Power & Mee, 2019), drop-in-centre as a space of care (Conradson, 2003) or verge gardens (Hsu, 2019). These non-animate materialities are therefore vital to the ways in which care is mediated with and through CFIs (Williams, 2020). Hsu (2019) in particular highlights how the presence of verge gardens provokes embodied practices of care for self and others. Paying attention to how more-than-human others enable care and care with diverse CFIs can assist us in understanding how CFIs are constituted in time and space.

Expanding understandings of the multiple others who constitute the communities of CFIs assists in facilitating further reflections on how diverse care needs and conflicting desires might be negotiated in unequal worlds. Beacham (2018) provides the example of such conflicts and reminds us of the limits of non-human agency. In his study of a CSA called Flourishing Fare, Beacham (2018) discusses a persistent beetle pest that was damaging the leafy green vegetables that were going to be sold to support the financial sustainability of the CFI. The beetle infestation was so severe that the group decided to temporarily use a pesticide to kill the beetle and stop it from further devastating the crop. Such an example reveals the complexity of care and the ways in which CFIs have to negotiate competing interests such as the life of pests versus the financial sustainability of the CFI. Actors involved in CFIs are therefore regularly confronted with choices to care for one aspect over another which may mean valuing human life above that of non-human others (Beacham, 2018). Once again, such insights highlight the exclusions that take place through and with care, as people negotiate diverse needs and values in unequal worlds.

Researching Sydney's community food provisioning initiatives

Care is best understood through grounded and contextual examples that facilitate an engagement with the situated needs being addressed by the caring practices of CFIs (Raghuram, 2016; Williams, 2017a). In Australia, the numbers of people experiencing food insecurity has risen from 15% in 2017 to 28% in 2021 (Food Bank Australia, 2017, 2021). Charity organisations dealing with food insecurity play a crucial role in taking responsibility for care by addressing hunger through the provision of food relief. However, organisations report having to turn people away due to insufficient food and resources, with less than two in five charities feeling as though they are able to meet the needs of the people they assist (Food Bank Australia, 2019). As Barosh et al. (2014) document, fresh food in Australia is becoming increasingly expensive and comprising a greater portion of average weekly incomes. In addition, food waste remains a key environmental concern, with approximately 7.3 million tonnes of food waste generated throughout the food production and consumption chain in 2016–2017 (ARCADIS, 2019).

Alongside these concerns, the impacts of conventional agriculture on the Australian environment and the long-term effects of climate change on food add pressure to a precarious food system (Hughes et al., 2015). The unfair wages paid to farmers, the concentration of farm production in the hands of multinationals, and the continued environmental impacts of conventional food production remain ongoing concerns affecting Australia's food system (Edwards, 2011, 2016; Larder et al., 2014, p. 57). Sydney, Australia's most populated capital, is heavily reliant upon the industrial agriculture system which was violently privatised from Aboriginal peoples despite ongoing resistance (Gaynor, 2018). In his study of food politics and dispossession in Australia, Mayes (2018) emphasises the ways in which cultivating lands for agriculture facilitated colonisation. He argues that the Australian local food and food sovereignty movements have the propensity to "depoliticise and mask historical injustices associated with colonialism" (Mayes, 2018, p. 13). Indeed, Mayes (2018) highlights the importance of making hidden injustices visible.

This chapter draws on research from a project that sought to understand the multiplicity of organisations responding to concerns around food insecurity, food waste, food justice, and sustainability in Sydney, Australia[1] in 2019 (see also Williams & Tait, 2022). Sydney is located on the lands of the Dharug, Eora, Gandangura, and Tharawal nations, whose cultures and customs have nurtured and continue to nurture these lands since the Dreamtime. The research was approved by the Macquarie University Human Research Ethics Committee and specifically focused on food provisioning CFIs in order to explore the inter-relationship between initiatives and performatively group these initiatives together as a sector responding to food system crisis. For the purposes of this chapter, we draw on two of ten semi-structured interviews with CFI coordinators. Interviewees were asked questions about the role and aims of their organisations, ways the organisations were resourced, how care was practiced through the initiatives, challenges they faced, and ways they might be overcome. The analysis of primary data was supplemented with selective website and social media analysis to provide further contextual information about the organisations studied and their activities.

Hopes and troubles of CFIs in Sydney

In what follows, we draw on the example of the ARCCO Food Pantry to reveal how flows of care can be traced through the relationality and connections that constitute the organisation through direct and mediated relationships. We discuss how care is practiced in response to unequal worlds of abundance and scarcity and the limits of this care evident in the landscape of food waste redistribution. Second, we draw on the example of Five Serves Produce CSA to attend to how this model of CFI is reliant upon human and more-than-human others who are made visible through care-thinking as actors that are integral to care through CFIs. Such insights reveal the troubles and hopes of CFIs that are held in tension, shaping their caring capacities in unequal worlds.

Addison Road Community Centre Organisation (ARCCO) Food Pantry: relational care in unequal worlds

The ARCCO Food Pantry fills a crucial role in addressing food insecurity and redistributing abundant food waste in the inner west of Sydney. This example reveals the ways in which CFIs operate with expansive understandings of community and dependence upon direct and mediated flows of care. Throughout this section, we highlight how needs for care were recognised and responded to by the ARCCO whilst highlighting the unequal and troubling contexts of abundance and scarcity that shape the work of food relief and food waste reduction. Our intent is not to romanticise care or celebrate the generosity that such accounts highlight, but to reflect upon what is revealed about the role of care in the work of CFIs.

The nine-acre block ARCCO occupies was handed over to the public in 1976 for community and recreational purposes (Addison Road Community Centre, 2020a). Overwhelmingly, the local community wanted to use the space to celebrate, maintain, and share cultures brought to Australia from around the world. It has since evolved into Australia's largest non-profit community centre and is deeply committed to "fighting for social justice, supporting arts and culture and caring for our environment" (Addison Road Community Centre, 2020b). Currently, ARCCO hosts 43 not-for-profit community development organisations and runs a number of programmes focused on nurturing community connection, solidarity, creativity and sustainability (Addison Road Community Centre, 2020c). The ARCCO plays a significant role as an advocate against racism and injustice for Aboriginal and Torres Strait Islander Peoples, migrants, and refugees.

In 2016, ARCCO recognised the increasing number of people experiencing food insecurity and established a food pantry with a small grant from the New South Wales Environmental Protection Authority (NSW EPA). The pantry rescues fresh fruit, vegetables, bread, and other high-quality food from being thrown into landfills and on-sells them "at about 40% off the retail value" (Interview ARCCO). Since it began, the food pantry has grown significantly from a 20-square-metre shipping container serving around 70 customers a week to a 100-square-metre building that caters to 1,500–1,800 individuals per week (Smart, 2019). With growing numbers of people accessing the pantry, ARCCO has had to respond and become increasingly sophisticated in their operations (Interview ARCCO). ARCCO food programmes have expanded to encompass mobile food pantries and Vege-table Mondays, a partnership with Foodbank,[2] Australia's largest food redistributor, in which 400–600 kg fresh fruit and vegetables are rescued from the Flemington (wholesale) markets and on-sold for $AUD2 per bag. In 2019, the Food Pantry was open Tuesday to Friday, with estimates that they sold about $AUD 20–25,000 worth of food per week to a range of people, including those concerned about food waste, people experiencing food insecurity, and "savvy shoppers" (Interview, ARCCO).

The care practiced by ARCCO is direct and involves embodied connections with known others who shop at or receive food hampers from the food pantry. The community cared for by ARCCO is dynamic and changes as need increases. By physically meeting the needs of people who are food insecure, shoppers are cared for by the organisation and are active in their receipt of care. The purpose of the food pantry is more than to provision food; it is also to support people and "create community" connection:

> We get the whole range here and we do that deliberately because . . . what we recognise [is] that first and foremost we are a community service . . . what we recognise is that for people who are experiencing food insecurity and poverty, a lot of them are actually excluded from the community and so by having an affordable supermarket that, isn't just about servicing poor people – or supporting poor people, we're actually creating a community and . . . that's much more reflective of the broader community. So, it's a pathway for people as well

> to become part of the broader community and to see themselves as part of a broader community. So, they're not just. . . . We're not just helping them with food, we're actually also building their resilience and their sense of connection with the local community.
>
> (Interview, ARCCO)

The pantry has a range of shoppers from those that are food conscious to those that are food insecure. Those cared for, with and through the food pantry directly include:

> some of the most vulnerable people in our local community, including people who are experiencing homelessness, unemployment, as well as students and others who find themselves struggling to put enough food on the table.
>
> (Addison Road Community Centre, 2020d)

In addition, Mordue (2021) reports that ARCCO supplied over 16,500 food hampers to over 160 groups throughout 2020. During the beginning of the COVID-19 crisis, they reported an increase in the number of people accessing the service to around 2,000, illuminating the effects of the public health lockdown measures that had been put in place to reduce the spread of the virus (Mordue, 2021). However, international students, asylum seekers, and refugees were excluded from government support during the COVID-19 pandemic, leading to additional people seeking assistance (Mordue, 2020).

Volunteers and staff play a role in enabling ARCCO to care for people who use the food pantry. At the time of the interview in 2019, before the COVID-19 pandemic increased focus on the work of ARCCO, they were finding it difficult to find enough volunteers, yet those they had were stable. However, in 2020, ARCCO Food Pantry saw a growth in the number of people volunteering from 62 at the beginning to 196 active by the end of the year and over 500 people across the year (Mordue, 2021). The existence of initiatives such as the ARCCO Food Pantry and their need to respond to growing levels of food insecurity make visible the unequal worlds and needs for care that exist.

The care practiced through ARCCO can be mediated and is reliant on interdependent relationships with food donors. ARCCO receives food from a range of sources including buying their own food, purchasing redistributed food from Foodbank Australia, and receiving food directly from donors. Donors practice mediated care, providing ARCCO with food that is essential to their care work, at the same time as limiting the impacts of food waste for donors whose food was slated for landfill. On their website, ARCCO lists 62 different groups that support them as food donors with 60,301 kg of food rescued between July 2019 and February 2020 (Addison Road Community Centre, 2020d). A significant amount of labour goes into engaging with the food donors, with a dedicated donor engagement officer working to understand why businesses do or do not and how to "overcome those barriers" (Interview, ARCCO). The variation in food donations poses some challenges for the organisation to develop a pricing model that is able to respond to these variations in supply, stay afloat, and care for shoppers:

> A challenging thing for us is about pricing the food. And to maintain some consistency because people don't like coming in and something they'd got last week was $2 and today it's $3. Because a dollar's a big thing for a lot of people . . . We're going through a stage of working out what our margins need to be . . . if we get food for free as opposed to if we've got to buy it from Foodbank or buy it from somewhere else . . . Something that we've really got to work on and figure out how to do that in a way that keeps us afloat . . . covering some of our overheads, but also make sure that the service we're providing remains affordable to the most needy groups that we're trying to reach.
>
> (Interview, ARCCO)

The organisation dedicates a significant amount of labour to promoting their cause through social media and their website, which includes advocacy campaigns dedicated to raising the rate of government support payments, which remain below the poverty line, and campaigns against racism. These media outlets also highlight their work in engaging with various food donors, who are core to their efforts to redistribute food waste.

Food waste is a by-product of overproduction and an abundance of food, whilst food insecurity reveals a multitude of interconnected and contextual inequalities that limit people's ability to access and afford food. Such examples reveal both the ongoing troubles facing people who need to access more affordable food staples through organisations such as the ARCCO food pantry, and the creative and dynamic responses developed by organisations with the capacity to care in unequal worlds. In 2018, ARCCO received a grant from the New South Wales Environmental Protection Agency to coordinate the *WOW Food! Inner West project*. This project "aims to reduce food waste going to landfill, and use this to help local people overcome food insecurity" (Addison Road Community Centre, 2020d). ARCCO explains that

> Food waste is a major issue in the Inner West. There are local families experiencing hunger today as tonnes of perfectly good food is being thrown out. Our "War on Waste (WOW) Food! Inner West" campaign is all about fighting food waste in the Inner West and making healthy food more accessible to people experiencing food insecurity. We are diverting tonnes of good food from going to landfill.
>
> (Addison Road Community Centre, 2020d)

The EPA grant is restrictive in that it is tied to alleviating food wastage and the environmental problems caused by excess food wastage. However, ARCCO has also used it to help establish an inner west Food Rescue Alliance, which enables organisations to mutually support and benefit one another, with the ARCCO website explaining how,

> Everyone knows we work better together so that is why we are connecting up with the many community service organisations across the Inner West who provide daily food relief for individuals and families. The Inner West Food Rescue Alliance, set up by Addi Road, will determine ways that we can all work together to recover more surplus food, share equipment and resources and reach out to more people in need. As it develops, the Alliance will provide even greater food security in our community.
>
> (Addison Road Community Centre, 2020d)

Such activities reveal the importance of caring capacities. ARCCO is able to expand the subjects and objects of care to those more-than-human others who are directly and indirectly affected by the greenhouse gas emissions caused by food waste.

The example of the ARCCO food pantry provides insight into how CFIs work with their existing capacities to care in unequal worlds. Staff, volunteers, users, donors, and alliance members and their networks are connected to ARCCO through relationships. Care is enacted both directly through the provisioning of food and indirectly through donations. The reliance of the ARCCO food pantry upon food waste redistribution highlights the tensions that pervade the landscape of food relief that is reliant upon the surplus production of food to feed those who are hungry (Isola & Laiho, 2020; Saxena & Tornaghi, 2018). At the same time, it reveals the way people are able to be cared for by such initiatives and the necessary care carried out in unequal worlds.

Five Serves Produce: more-than-human care in unequal worlds

In this section, we provide an example that further illuminates the ways in which care might be practiced by/for/with more-than-human others. Throughout this section, we highlight the labour and tensions involved in sustaining a CSA whilst caring for non-human and human others in unequal worlds, and by doing so, reflect upon what is revealed about the role of care in the work of CFIs. The example reveals the importance of communities that are cared for by CFIs and the ways in which more-than-human others become subjects and objects of care. At the same time, the tensions of such care are made apparent as people are constrained in their capacity to care due to non-human agencies and unequal worlds.

Five Serves Produce is a Community Support Agriculture Scheme (CSA) based in the North West of Sydney, Australia, and is one of the only CSAs in the Sydney area. In 2016, it began to organically supply food with twice-weekly deliveries and box pick-ups. Located on four acres of flat land on the banks of Dyarubbin (Dharug Aboriginal term for Hawkesbury/Nepean River), in Richmond. In 2019 it was located on four acres, Five Serves seeks to "steward and care for the land on which we grow our produce" (Five Serves Produce, 2021). The land is owned by the

Wheen Bee foundation that researches and conducts activities to support bees and food security. Five Serves delivers organic food boxes in two different sizes to the Western suburbs of Sydney and Blue Mountains regions. Whilst not certified the farm "[does] everything organically" (Interview, Five Serves Produce) and is reliant upon direct relationships between members and the farmer where relationships of trust means that independent certification is not necessary. Five Serves Produce is reliant on interdependent relationships between members and farmers who develop reciprocal relationships of care and responsibility:

> I still call them members, not customers, because I want them to know they have input and they can ask me to grow more of something, less of something, and they share the risk as well you know. If something fails they have to not have it.
>
> (Interview, Five Serves Produce)

Participation in the CSA by members enables them to take responsibility for caring through the CSA. Running the CSA is hard work and demands a significant time and energy commitment from the manager who founded and continues to operate the farm.

Five Serves Produce cares for and is cared for by staff, members, and non-human others who constitute the CFI through their caring practices. In 2019, Five Serves Produce directly supported four staff through award-wage employment and supplied 75 boxes per week. Unlike other CSAs, there are no volunteer requirements as members support the farm through their subscription payments rather than their labour, except for occasional working bees. Members purchase 6 or 12 weeks' worth of produce, which they may have delivered or pick up from the farm each week. Subscribers become part of the CFI community, being cared for through their subscription to healthy, organic, weekly vegetable boxes that are filled with fresh seasonal vegetables and fruit. CSA schemes can be shaped by a diversity of motivations, with Wells and Gradwell (2001, p. 113) finding a range of caring practices enacted in their study of CSAs including care for non-human others, people, food, education, community, and future wellbeing. Retaining membership numbers is an ongoing struggle as transient memberships comprised 2/3 or box subscription in 2019:

> It's getting better, but there are a lot of people who try it and don't like it. We probably only have a third of our members that have been with us for longer than a year. There's another third who were kind of – couple of subscriptions in, and then there's a third who just – they'll do one subscription, and they'll leave . . . We're not anywhere over . . . sixty percent retention rate. I'd be happy with that if I could get it.
>
> (Interview, Five Serves Produce)

Being reliant on seasonal crops means members have to accept the variations in what is in season, crop failures, disruptions caused by flooding, or pests and potentially adjust their cooking practices to use the food that is able to be grown. Five Serves continues to market to new members: to encourage people to care about the goals of the CSA and heighten awareness of the benefits of organic farming for humans and non-humans. Farm tours are offered for people to learn about how the food is grown, along with other picnics for socialisation between CSA members at particular points in time.

The membership composition of the CSA is however dynamic, with some people finding the local food box model does not work for them whilst others "can't afford to continue" (Interview, Five Serves Produce) due to unexpected bill payments or living costs. Like other CSA subscriptions (Sherriff, 2009), the cost of being a member can be prohibitive for low-income groups, despite prices being less than many online organic food suppliers. Such examples reveal the tensions inherent in farming organically through CSA models as they get established:

> in an ideal world, if we're making good profit, I'd love to incorporate that [low-income price] somehow whether it's with . . . a scholarship voucher type system . . . my other members can all contribute an extra $5.00 and then we can give a box to a needy family each week. But that's like in the five to ten-year plan. When I'm.
>
> *Interviewer:* Able to pay yourself?
>
> . . . Yes, a living wage; and then I can do that.
>
> *(Interview, Five Serves Produce)*

People's capacity to care is restricted by unequal worlds that make organic produce financially affordable for some but not others due to complex structural factors. It is necessary and vital for CSA producers to "make a living" (Wells & Gradwell, 2001, p. 113) in order to survive. Whilst the CSA remains financially marginal, the capacity for care beyond members and non-human others is restricted.

Five Serves Produce is dependent on a diverse community of non-human others who care with the CSA, which is reflected in its aims. An expansive community comprising local people, the land, soil, flora, fauna, and the river is included through relational connections of benefit and the reduction of harm by the CSA:

> to be a welcoming place where the local community can meet their farmer and learn more about how their food is grown. . . . Another equally important aim of our farm is to steward and care for the land on which we grow our produce and to reduce food miles. . . . Food is produced in an ecologically responsible way and only organic farming practices are used. This ensures that the soil microbes, the flora and fauna in the river nearby and the local community who eat the food, benefit from the existence of the farm and are not harmed by it. . . . We try to reduce the use of plastic wherever possible and minimise packaging of our vegetables.
>
> (Interview, Five Serves Produce)

By farming organically, they rely on working with natural systems and "using people power to do the work" rather than "pesticides and herbicides" (Interview, Five Serves), which are more labour-intensive, but essential to how they care for non-human others. Further practices including on-site composting, organic farming methods, minimising plastic use on the farm and reducing food waste by members are also apparent:

> We'll do things like put the broccoli leaves in with the broccoli instead of cutting them off, and tell people they can eat them . . . and then saying to people, when we have excess, we had some pumpkins that were slightly damaged and we just put a box out at the pickup time, and the pickup customers can just take extra. We've also put out a thing in the newsletter asking who's interested in wonky veg and so some people responded. They get all the weird carrots, as well as their bunch of carrots so they're getting extra veggies. . . . I've done lots of posts on Instagram and Facebook with a kale leaf with a hole, just saying "It's so good even the slugs want to eat it."
>
> (Interview, Five Serves Produce)

Overwhelmingly, it is the CSA that bears the risks that come with farming, choosing to prioritise caring for members over their own income, particularly in response to adverse weather events:

> for a month in February this year (2019) nothing would grow, everything flowered, the tomatoes stopped growing, stuff burnt. We just paused everyone's subscriptions and we lost income for that month but they didn't lose out. They just went on a break and then we started up again once we had the stuff coming through. So, we need to put up shade cloth next year.
>
> (Interview, Five Serves Produce)

Like the example of the persistent beetle pest provided by Beacham (2018), CSAs are often negotiating being financially sustainable and working with the non-human world, which asserts its agency to reduce their impact, grow organically, and minimise harm. In this case, the manager of Five Serves notes the effects of extreme heat on the ability to grow food and the ways in which they sort to respond and adapt by placing shade cloth the following year.

Tracing these multiple forms of contingent and relational care through Five Serves reveals the ways in which they are constituted through direct connections between the CSA, soils, microbes, non-human others, and members. The farm itself is a place that provides care and is cared for through the CSA model, which highlights the interdependence between farmers, members, soils, seasons, and crops. The community that is dynamic and temporal is constantly being reconfigured through their connections to the CSA. Transactions of membership and labour signal who becomes the community that is cared for and by Five Serves Produce. At the same time, the microbes, soils, worms, and fruit and vegetables care with, through, and for the CFI. Some, like the bunch of unusually shaped carrots above, facilitate care for others fleetingly, while others, like the land and bees, are cared for in perpetuity. Bodies are nourished and livelihoods are provided for as long as the CSA can encourage enough people to care and become part of its community.

Conclusion

This chapter has examined two different forms of CFIs and reflected on how paying attention to care might help us understand the hopes and troubles that affect CFIs in unequal worlds. Paying attention to care as a vital practice constituting communities of CFIs reveals the expansive communities that are cared for, cared about and care with CFIs. The subjects and objects of care differ depending on the goals and aims of the CFI, which may be focused on mitigating the impacts of food waste, responding to food insecurity, caring for the environment through sustainable farming practices, and caring for the health of members through organic food.

The very need for types of care through food relief is evidence of the brokenness of food systems (Williams & Tait, 2022) and reflects how actors seek to make "as well as possible worlds" (Puig de la Bellacasa, 2017, p. 7) in the here and now. Caring is hard work, relentless, responsive, and never finished. In line with a critical-reparative approach, we have acknowledged the potentiality of care enacted with, through and by CFIs to "maintain, continue and repair our worlds" (Fisher and Tronto cited by Tronto, 1993, p. 113). We have also acknowledged the exclusions and limitations of who and what is able to be a subject of care and the tensions inherent in different models of CFI due to resourcing, funding structures, and the constraints of caring capacity in unequal worlds.

As both the literature and examples have shown, when paying attention to the expansive communities that are enrolled in relational caring, we take note of the presence of more-than-human others in such communities. More-than-human others may be the subjects and objects of care (Power & Williams, 2019), being cared for and mediating care for people involved in CFIs. Paying attention to caring practices also reveals the ethical negotiations that take place about what is or is not cared for due to the capacities of different forms of CFIs, with CSAs more able to take care of soils and worms than food relief initiatives whose primary concern is feeding people and reducing food waste. Materials and objects, such as food itself, nourish bodies, caring for members and mediating the care that flows through ways of "doing" CFIs. Caring for more-than-human others involves work, and as revealed in the CSA example, can be more labour intensive than less caring methods of farming such as using chemicals and pesticides. In discussing the limits on caring capacities that are created in unequal worlds, we are able to see how the communities that form CFIs are restrictive or expansive and the limits of food waste redistribution or subscription-based CSA systems. At the same time, we note the very real and important work these CFIs are doing in responding every day to the need for care, with care, and the insights that can be garnered about the hope and trouble of CFIs by paying attention to care in unequal worlds.

Acknowledgements

We would like to thank the research participants for their contribution to the project. The project was funded through a Macquarie University New Staff Grant and Early Career Researcher Fellowship Award from Macquarie University. Thank you to the book editors and anonymous reviewers for their helpful suggestions.

Notes

1 The research design was approved by the Macquarie University Human Research Ethics Committee and funded through a Macquarie University New Staff Grant in 2019–2020.
2 Food Bank Australia is a not-for-profit organisation that redistributes excess food and donated food from food manufacturers, growers, producers, retailers, and wholesalers to charities. It does not provide food directly to individuals. Food bank provisions both fresh fruit and vegetables and grocery items. The majority of food relief charities in Australia will be reliant on food bought or provisioned for free by Food Bank Australia.

References

Abbots, E. J., Lavis, A., & Attala, L. (2015). *Careful eating: Bodies, food and care*. Routledge.

Addison Road Community Centre (2020a). *About us*. https://addiroad.org.au/about-us/

Addison Road Community Centre (2020b). *Addison road community centre*. https://addiroad.org.au/

Addison Road Community Centre (2020c). *Our tenants*. https://addiroad.org.au/our-tenants/

Addison Road Community Centre (2020d). *WOW food war on waste*. https://addiroad.org.au/wow-food-innerwest/

Alam, A., & Houston, D. (2020). Rethinking care as alternate infrastructure. *Cities*, *100*, 1–10. https://doi.org/10.1016/j.cities.2020.102662

Alkon, A. H., & Guthman, J. (2017). Introduction. In A. H. Alkon & J. Guthman (Eds.), *The new food activism opposition, cooperation, and collective action* (pp. 1–27). University of California Press.

ARCADIS (2019). *National food waste baseline final assessment report executive summary*. https://www.environment.gov.au/system/files/pages/25e36a8c-3a9c-487c-a9cb-66ec15ba61d0/files/national-food-waste-baseline-executive-summary.pdf. Accessed 26 February 2020.

Barosh, L., Friel, S., Engelhardt, K., & Chan, L. (2014). The cost of a healthy and sustainable diet – who can afford it? *Australian and New Zealand Journal of Public Health*, *38*(1), 7–12. 10.1111/1753-6405.12158

Bawaka Country, W. S., Suchet-Pearson, S., Lloyd, K., Burarrwanga, L., Ganambarr, R., Ganambarr-Stubbs, M., Ganambarr, B., Maymuru, D., & Sweeney, J. (2016). Co-becoming Bawaka: Towards a relational understanding of place/space. *Progress in Human Geography*, 40, 455–475. https://doi.org/10.1177/0309132515589437

Beacham, J. (2018). Organising food differently: Towards a more-than-human ethics of care for the anthropocene. *Organization*, *25*(4), 533–549. https://doi.org/10.1177/1350508418777893

Bedore, M. (2018). "I was purchasing it; it wasn't given to me": Food project patronage and the geography of dignity work. *The Geographical Journal*, *2018*(184), 218–228.

Cloke, P., May, J., & Williams, A. (2017). The geographies of food banks in the meantime. *Progress in Human Geography*, *41*(6), 703–726. https://doi.org/10.1177/0309132516655881

Conradson, D. (2003). Spaces of care in the city: The place of a community drop-in centre. *Social & Cultural Geography*, 4(4), 507–525.

Cox, R. (2010). Some problems and possibilities of caring. *Ethics, Place & Environment*, *13*(2), 113–130.

Dowler, E., Kneafsey, M., Cox, R., & Holloway, L. (2010). 'Doing food differently': Reconnecting biological and social relationships through care for food. *The Sociological Review*, 57, 200–221.

Edwards, F. (2011). Small, slow and shared: Emerging social innovations in urban Australian foodscapes. *Australian Humanities Review*, 51, 115–134.

Edwards, F. (2016). Urban food security and alternative economic practices. In R. Horne, J. Fein, B. B. Beza, & A. Nelson (Eds.), *Sustainability citizenship in cities* (pp. 40–51). Routledge.

Five Serves Produce (2021). *About*. https://fiveservesproduce.com.au/pages/about

Food Bank Australia (2017). *Foodbank hunger report 2017*. www.foodbank.org.au/wp-content/uploads/2017/10/Foodbank-Hunger-Report-2017.pdf

Food Bank Australia (2019). *Foodbank hunger report 2019*. https://www.foodbank.org.au/wp-content/uploads/2019/10/Foodbank-Hunger-Report-2019.pdf?state=nsw-act

Food Bank Australia (2021). *Food bank hunger report 2021*. https://reports.foodbank.org.au/wp-content/uploads/documents/2021-Foodbank-Hunger-Report-PDF.pdf

Gaynor, A. (2018). Learning from our productive past. In N. Rose & A. Gaynor (Eds.), *Reclaiming the urban commons: The past, present and future of food growing in Australian towns and cities* (pp. 167–174). Univeristy of Western Australia Publishing.

Gibson-Graham, J. K. (2006). *A postcapitalist politics*. University of Minnesota Press.

Goodman, M. K., Maye, D., & Holloway, L. (2010). Ethical foodscapes?: Premises, promises, and possibilities. *Environment and Planning A*, *42*, 1782–1796.

Hsu, J. (2019). Public pedagogies of edible verge gardens: Cultivating streetscapes of care. *Policy Futures in Education*, *17*(7), 821–843.

Hughes, L., Steffen, W., Rice, M., & Pearce, A. (2015). Feeding a hungry nation: Climate change, food and farming in Australia. *Climate Council*. https://www.climatecouncil.org.au/uploads/7579c324216d1e76e8a50095aac45d66.pdf

Isola, A.-M., & Laiho, J. (2020). Commoning surplus food in Finland – actors and tensions. In T. Eskelinen, T. Hirvilammi, & J. Venalainen (Eds.), *Enacting community economies within a welfare state* (pp. 95–116). Mayfly Books.

Johnsen, S., Cloke, P., & May, J. (2005). Day centres for homeless people: Spaces of care or fear? *Social & Cultural Geography*, 6(6), 787–811.

Kneafsey, M., Holloway, L., Venn, L., Cox, R., Dowler, E., & Tuomainen, H. (2008). *Reconnecting producers, consumers and food: Exploring alternatives*. Berg.

Larder, N., Lyons, K., & Woolcock, G. (2014). Enacting food sovereignty: Values and meanings in the act of domestic food production in urban Australia. *Local Environment*, *19*(1), 56–76.

Lawson, V. (2007). Geographies of care and responsibility. *Annals of the Association of American Geographers*, 97(1), 1–11.

Mayes, C. (2018). *Unsettling food politics: Agriculture, dispossession and sovereignty in Australia*. Rowman & Littlefield.

McEwan, C., & Goodman, M., K. (2010). Place geography and the ethics of care: Introductory remarks on the geographies of ethics, responsibility and care. *Ethics, Place & Environment, 13*(2), 103–112.

Mordue, M. (2020, March 31). *Feeding the people – Addi road, Craig Foster and Anthony Albanese*. https://addiroad.org.au/news/feeding-the-people-addi-road-craig-foster-and-anthony-albanese/

Mordue, M. (2021). *Food security, social justice and Donna Summer*. https://addiroad.org.au/news/food-security-social-justice-and-donna-summer/

Morrow, O. (2019). Sharing food and risk in Berlin's urban food commons. *Geoforum, Early View*, 1–11.

Morrow, O., & Martin, D. G. (2019). Unbundling property in Boston's urban food commons. *Urban Geography*, 1–21. https://doi.org/10.1080/02723638.2019.1615819

Nancy, J. L. (1991). Of being-in-common. In Miami Theory Collective (Ed.), *Community at loose ends*. University of Minnesota Press.

Popke, J. (2003). Poststructuralist ethics, subjectivity, responsibility and the space of community. *Progress in Human Geography, 27*(3), 298–316.

Power, E. (2019). Assembling the capacity to care: Caring-with precarious housing. *Transactions of the Institute of British Geographers, 44*, 763–777. https://doi.org/10.1111/tran.12306

Power, E., & Mee, K. (2019). Housing: An infrastructure of care. *Housing Studies, 35*(3), 484–505. https://doi.org/10.1080/02673037.2019.1612038

Power, E., & Williams, M. (2019). Cities of care: A platform for urban geographical care research. *Geography Compass, 14*(1), 1–11. https://doi.org/10.1111/gec3.12474

Puig de la Bellacasa, M. (2010). Ethical doings in nature cultures. *Ethics, Place & Environment, 13*(2), 151–169. https://doi.org/10.1080/13668791003778834

Puig de la Bellacasa, M. (2012). "Nothing comes without its world": Thinking with care. *The Sociological Review, 60*(2), 197–216. https://doi.org/10.1111/j.1467-954X.2012.02070.x

Puig de la Bellacasa, M. (2017). *Matters of care: Speculative ethics in more than human worlds*. University of Minnesota Press.

Raghuram, P. (2016). Locating care ethics beyond the global north. *ACME: An International Journal for Critical Geographers, 15*(3), 511–533.

Saxena, L. P., & Tornaghi, C. (2018). *The emergence of social supermarkets in Britain: Food poverty, Food waste and Austerity retail*. Coventry University.

Sharp, E. L. (2019). Editorial: The role of reflexivity in care-full food systems transformations. *Policy Futures in Education, 17*(7), 761–769. https://doi.org/10.1177/1478210319874256

Sherriff, G. (2009). Towards healthy local food: Issues in achieving just sustainability. *Local Environment, 14*(1), 73–92.

Smart, A. (2019, August 14). Food, affordable food. *City News*. http://cityhubsydney.com.au/2019/08/food-affordable-food/142196/

Tronto, J. (1993). *Moral boundaries, a political argument for an ethic of care*. Routledge.

Tronto, J. (2013). *Caring democracy: Markets, equality, and justice*. New York University Press.

Tynan, L. (2021). What is relationality? Indigenous knowledges, practices and responsibilities with kin. *Cultural Geographies, 28*(4), 597–610.

Wells, B. L., & Gradwell, S. (2001). Gender and resource management: Community supported agriculture as caring-practice. *Agriculture and Human Values, 18*(1), 107–119.

Williams, M. (2017a). Care-full justice in the city. *Antipode, 49*(3), 821–839. https://doi.org/10.1111/anti.12279

Williams, M. (2017b). Searching for actually existing urban justice. *Urban Studies, 54*(10), 2217–2231. https://doi.org/10.1177/0042098016647336

Williams, M. J. (2020, March). The possibility of care-full cities. *Cities*, 98, 1–7. https://doi.org/10.1016/j.cities.2019.102591

Williams, M. J., & Tait, L. (2022). Diverse infrastructures of care: Community food provisioning in Sydney. *Social & Cultural Geography*, 1–21. https://doi.org/10.1080/14649365.2022.2056630

Wilson, S. (2008). *Research is ceremony: Indigenous research methods*. Fernwood Publishing.

3 Understanding vulnerability and resilience of urban food initiatives in Morocco

Patrizia Pugliese, Cosimo Rota, Fatima Zohra Sabrane, Marie Reine Bteich, and Esther Veen

Introduction: a critical reparative approach towards urban agriculture

Inherently and imaginatively dichotomous in its very name, "urban" "agriculture" is often described through its multiple dualities, such as urban/rural, urban/peri-urban, traditional/innovative, risky/healthy, local/global, subsistence/professional, north/south, and hope/trouble. Rather than being rigid divisions, however, on the ground, these *clean conjectures meet messy reality* (Davidson, 2017, p. 68). Operating in grey areas, by nature at the intersection of spatial and organisational terms, the fuzziness and ambiguity of urban agriculture is best understood when its hybridity is acknowledged: this hybridity both hides and nurtures hopes as well as trouble. In addition, urban agriculture shows that it belongs to different sectors and domains of intervention.

With all this hybridity and diversity, urban agriculture is a complex, kaleidoscopic phenomenon. Over the past decades, it has drawn the attention of scholars from different disciplines and strands of thought. A growing body of academic research, extensively complemented by empirical evidence from various networks and projects, has described urban agriculture's distinct features and development trajectories in different regions of the world. Urban agriculture's diverse forms (home subsistence, family-type commercial farming, entrepreneurial farms (De Bon et al., 2010)) and multiple functions (food and nutritional security, poverty alleviation, social inclusion, agro-ecological management, environmental education, and recreation (De Zeeuw et al., 2011; Zasada, 2011)) have been classified, and the wide array of concerned actors has been mapped. Common issues of concern for the future of urban agriculture range from production limits and challenges around food safety and environmental pollution to access to land and labour or institutional fragility (Davidson, 2017). In some contexts, urban agriculture urges for renewal, and actualisation of concepts and practices (Moustier, 2017; Lavergne, 2004). At the same time, urban agriculture's adaptive capacity is widely praised, considering that it has proven to resist marginality and opposition, even in difficult environments. Reasons for encouragement focus on the responses that urban agriculture can provide to urban dwellers' changing needs and the role it can play in transitions to sustainable (urban) food systems (Valette et al., 2020).

DOI: 10.4324/9781003195085-4

Fragile but adaptive, agriculture in urban and peri-urban contexts is the object of different troubling pressures as well as hopeful initiatives. This dynamic interplay of visions, powers, and actors makes urban agriculture an interesting arena to reflect on the critical reparative approach proposed by this book. In this chapter, the critical reparative stance is articulated within the boundaries and with the vocabulary of an adapted version of Turner's vulnerability – and resilience – assessment framework (Turner et al., 2003; Knapp et al., 2016), methodologically integrating the findings of a main quantitative survey with crossing-over pre- and post-survey qualitative data (Datta, 2001).

We take this critical reparative stance, focusing on both the vulnerability – as a source of trouble – and the resilience – as a source of hope – of four cases of urban agriculture in Morocco. All four cases represent hybrid, fluid experiments operating within different networks. We consider these cases as community food initiatives as they are strongly societally engaged (for instance, by employing people with disabilities), harbour a clear element of improving the food system as we know it, try and create closer relations between producers and consumers, or address citizens' concerns through food discourse. While the four cases are not isolated eccentricities (Rosol, 2020; Sarmiento, 2017), they cannot (yet) be considered to form a niche that is transforming the present food regime. Rather, they should be seen as constituents of new, progressing assemblages (Marsden et al., 2018): they are places of cross-contamination, reflection, and reconfiguration, captured in the process of "becoming" seeds of transition. By introducing new concepts and practices, they are embedding hope in their own local contexts, where, however, unequal power relations trouble their subsistence and potential for transformative change.

With Sarmiento (2017, p. 486), we believe that initiatives such as these – vulnerable, but aiming for change – should be defined not only by their progress and success but also by their limitations and failures in their attempts at building up, consolidating, or diluting their alterity. Therefore, in this chapter, we advance a critical reparative approach through better understanding what makes the four urban agriculture initiatives vulnerable – that way highlighting their troubles – but also what makes them resilient towards those troubles – bringing out their hopes. By making both vulnerability and resilience a systematic area of inquiry, we provide a valuable opportunity to look at initiatives' struggles as well as their capacity for survival. Moreover, examining both vulnerabilities and resilience helps us promote a critical reparative stance by looking beyond stereotypes. One such stereotype of urban agriculture is an understanding of initiatives as *signs of the past* (Moustier, 2017, pp. 7, 17): residual subsistence agricultural patches surviving on degraded vacant soil, struggling often unrecognised within rapid urban sprawl. Another stereotype is a perspective on initiatives as innovative, forward-looking urban agriculture integrated in the *metabolisme urbaine* (Aubry & Porias, 2012, p. 152) believed to build *food bridges* and contribute to sustainable urban growth through multi-actor engagement within and beyond agriculture (Davidson, 2017, pp. 70, 72). Moving beyond such limiting classifications enables us to better capture hope and trouble in current hybridisation trends without neglecting what is happening in the middle.

In sum, the aim of this chapter is to understand the extent to which the four initiatives studied are susceptible to perturbations that feed their vulnerability. Also, we are interested in what strengthens their resilience. In the next section, we first introduce the context of urban agriculture in Morocco. After that, we present the vulnerability assessment framework, which we used to structure our analysis. Thereafter we discuss the methods we employed to study the four cases and present these cases in more detail, after which we present our findings, explaining the main vulnerabilities of the four cases and how they deal with them. We end with some overall conclusions.

Urban agriculture in Morocco

Urban agriculture scholarship generally differentiates between urban agriculture in the global north and urban agriculture in the global south (another, smaller body of literature discusses central and eastern Europe). This common geographic north/south divide, however, is progressively blurred. Urban agriculture initiatives in Northern developed countries used to be discussed in terms of their aesthetic, environmental, and urban-rural reconnection functions. Urban agriculture in Southern, developing contexts was dominantly understood to be driven by urban food security and poverty alleviation. This simplistic geographical division appears increasingly less appropriate. Indeed, a switch and subsequent convergence of priorities and discourses between the global north and south are observed. Stakes and concerns behind urban agriculture are changing (Aubry & Pourias, 2012), with nutritional and food security issues becoming crucial in the cities of developed countries, while food safety, social inclusion, and recreational-educational narratives gain attention and support in developing contexts. Clearly, assuming differences based on different contexts is too simplistic (Sovová & Veen, 2020). Subsequent alignment and convergence of research questions, methods, and tools cannot go unnoticed (Aubry & Porias, 2012).

Urban agriculture in Morocco brings together discourses stemming from the global North and South, underscoring hope and trouble. Agriculture has been a distinctive element of Mediterranean urban landscapes since ancient times (Lavergne, 2004), resisting the impact of industrialisation, agricultural modernisation, and exponential urbanisation (Jouve & Padilla, 2007; Soussan, 2015). Although Mediterranean urban agriculture remains substantially informal, being ignored by official statistics and neglected by agricultural and urban policies (Valette & Philifert, 2014), urban agriculture in Morocco – as in other Southern Mediterranean countries – plays a role in the ongoing debate about the modernisation of urban food distribution: it caters for hopes to alleviate food insecurity of low-income parts of the urban population as well as for the high classes' tastes for specific local traditional products (Rousseau et al., 2020). Supported by local associations, international donors, and members of the Moroccan diaspora, it also takes the shape of innovative agri-food initiatives. While unique in their origin and development trajectories, these initiatives share the ambition to propose a hopeful, alternative food discourse, combining health, social, and environmental concerns, as an increasing share of Moroccan citizens perceive these as not sufficiently considered by agricultural and urban policies. Initiatives present themselves as places of producing, selling, consuming and interacting around "good" foods, adopting organic or agroecological farming practices, piloting circular economy solutions, and offering job opportunities to vulnerable groups. Troubled to receive attention and support, they evolve in a grey area of action where they engage their target communities and public in diverse ways, through and beyond food connections (Davidson, 2017).

Vulnerability and resilience

The present work builds upon that of Knapp et al. (2016), who studied the resilience of 29 urban agriculture projects in Switzerland and the Netherlands. They adapted the vulnerability framework designed by Turner et al. (2003) to enable it to study vulnerability and resilience in urban agriculture (see Figure 3.1). Following Eakin and Wehbe (2009), Knapp et al. defined urban agriculture projects as coupled socio-ecological systems affected by both human influences – such as institutions and policies – and environmental aspects, like soil, water, sun, and pests. Their starting point was that projects face perturbations, to which they are vulnerable; their resilience helps them overcome these perturbations (Knapp et al., 2016).

Vulnerability is defined as the degree to which a system, subsystem, or system component is likely to experience harm due to exposure to a *hazard*. A hazard is a threat to a system that can comprise *perturbations* or *stress* and their consequences (Turner et al., 2003). A perturbation is a major spike in pressure beyond the normal range of variability in which the system operates (Turner et al., 2003). Perturbations usually originate beyond the system or the location in question – while stress

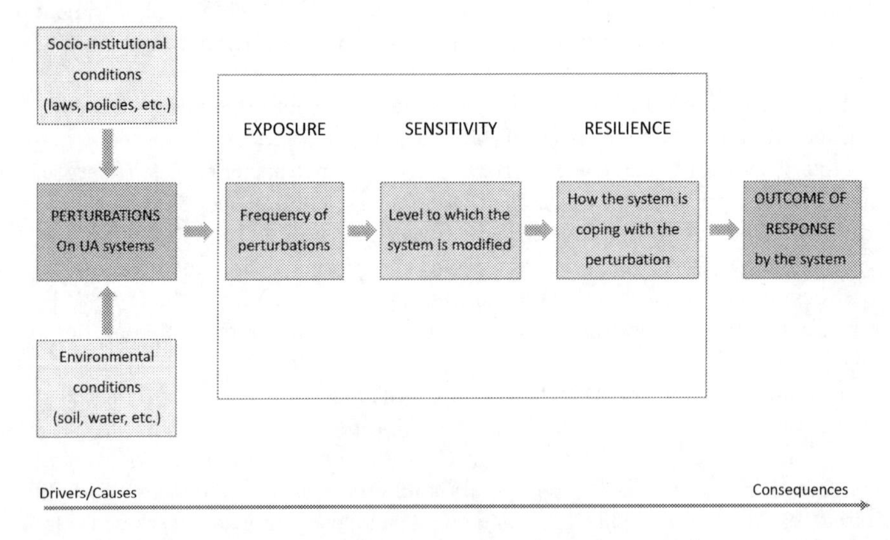

Figure 3.1 Vulnerability framework for urban agriculture projects (adapted from Turner et al. (2003) by Knapp et al. (2016))

generally resides within the system (Turner et al., 2003). In this chapter, we simply define a perturbation as a situation or occurrence that threatens the existence of an initiative. Perturbations can be internal or external, sudden or gradual, and their causes can lie in both human and environmental factors.

Vulnerability can be understood as a state of fragility or a disposition to be hurt (Urruty et al., 2016). Resilience, conversely, is the ability of a system to absorb shocks, recover from stress, and adapt to further perturbations (Urruty et al., 2016). Resilience therefore shows the potential viability and lifespan of an initiative (Knapp et al., 2016). The vulnerability assessment as used by Knapp et al. (2016) is based on the conceptual approach of Adger (2006), who relates vulnerability to the combined effects of (1) the *exposure* to perturbations, (2) the *sensitivity* to these events, and (3) the possibilities for *resilience*. Exposure in this view is defined as the frequency, intensity and duration of affecting perturbations (Urruty et al., 2016), whereas sensitivity is defined as the degree to which the system is affected by this exposure (Adger, 2006). Figure 3.1 shows this vulnerability framework graphically.

A project's vulnerability towards a certain perturbation thus depends on the interaction of exposure, sensitivity, and resilience. Commonly, when resilience to a perturbation is low and exposure and sensitivity are both high, the project is relatively vulnerable to this perturbation. Conversely, a high resilience may compensate for high exposure and sensitivity, resulting in the project being resilient or neutral towards this perturbation.

As stated, in this chapter, we treat vulnerability as a source of trouble and resilience as a source of hope. However, we acknowledge that vulnerability and resilience are inherently connected in a causal analytic loop. In an attempt to blur the simplified dichotomic thinking that is often present in urban agriculture writing, we stress that perturbations may sometimes represent seeds of positive change and sources of hope and that strategies to reinforce resilience are not all equally sustainable and thus hopeful. By acknowledging that reality is more complicated than connecting resilience with hope and vulnerability with trouble, we aim to reinforce the critical reparative approach as developed throughout this book.

Table 3.1 lists the specific challenges, pressures, and constraints that urban agriculture projects in the Mediterranean and developing countries face. We used two steps to create this list. The first step was a literature review performed in an iterative process. We collected both academic and non-academic literature (e.g. reports and policy documents), which helped us create an initial list of perturbations common to urban agriculture projects in Mediterranean countries in general and Morocco in particular. Second, we supplemented this list with perturbations mentioned by respondents representing Moroccan case studies (disclosed as "field outcome" in Table 3.1), leading to a list of 24 perturbations in total. These can be categorised into five groups: (1) project location and structural features, (2) project management, (3) interaction with market actors, (4) interaction with institutional settings, and (5) interactions with society. The majority of these perturbations are also found in other contexts, though nine perturbations seemed specific to Morocco and developing countries: water scarcity, bad quality of irrigation water, low profitability of production, climatic hazards, privatisation of land, dispensation law, high cost of organic certification, lack

Table 3.1 Perturbations for urban agriculture initiatives in developing countries with a focus on Morocco, based on literature and field work

Perturbation	*Description*	*Source*
1. Project location and structural features		
Difficult access to land	High rate of urbanisation; uncertainty of land availability; high land prices; urban sprawl	Benabed et al., 2014, De Bon et al., 2010, Debolini et al., 2015, Dugué et al., 2015b, Giseke et al., 2015, Moustier, 2017, Sarmiento, 2017, Valette et al., 2020
Insecure land tenure	Absence of secured tenure statues	Lorenz, 2015, Valette et al., 2020, Valette & Dugué, 2017
Soil degradation	Physical, chemical, and biological decline in soil quality	Moustier, 2017
Land contamination	Pollution of lands and its effect on public health	De Bon et al., 2010, Dugué & Valette, 2015, Lorenz, 2015, Moustier, 2017
Water scarcity	Lack of water in between seasons	Moustier, 2017, Sarmiento, 2017
Bad quality of irrigation water	Pollution of irrigated water by sewage	Debolini et al., 2015, Dugué et al., 2015a
Climatic hazards	Unpredicted climatic events (such as draught, hail, or floods)	Aubry & Pourias, 2012, Callo-Concha & Ewert, 2014, Davidson, 2017
2. Project management		
Difficult access to organic inputs	Difficult access to and reliance on external inputs: seeds, compost, animal manure, biopesticides, biofertilizers	De Bon et al., 2010, Lorenz, 2015
Difficult access to financial capital	Lack of financial resources and sponsors	Benabed et al., 2014, Orsini et al., 2013
Low profitability of production	Low profitability and discontinuous production	Debolini et al., 2015
Lack of labour availability	Strong competition for access to manpower	De Bon et al., 2010, Moustier, 2017
Crop diseases	Reduction in crop yield and quality	*Field outcome*
3. Interactions with market actors		
Undeveloped marketing pathways	Low level of organisation of farmers in the market; unlikeliness of direct sales; lack of short circuits	Banzo et al., 2016, Barbizan, 2011, Benabed et al., 2014, Dugué et al., 2015a, 2015b, Dugué & Valette, 2015, Moustier, 2017

Perturbation	*Description*	*Source*
Lack of marketing manager	Lack of dedicated marketing professionals	*Field outcome*
Lack of variety of product	Limited offer of product varieties	*Field outcome*
4. Interactions with institutional settings		
Dispensation law	Approval of building permits conflicting with municipal zoning plan and general master plan	De Zeeuw et al., 2011
Absence of supportive policies	Absence of policies promoting urban agriculture from territorial authorities	Banzo et al., 2016, Dugué et al., 2016
Privatisation of collective land	The conversion of collectively managed land in the Gharb irrigated perimeter in Morocco into private property as per the Morocco's Employability and Land Compact	Debolini et al., 2015
Lack of advisory services	Absence of contact between extension services and farmers	Giseke et al., 2015
Absence of policies	Lack of programmatic and development policies for urban agriculture	Benabed et al., 2014
High cost of organic certification	Difficult to get organically certified because of high costs	Lorenz, 2015
Administrative constraints	Excessively complicated administrative procedure	*Field outcome*
5. Interactions with society		
Neighbours' misbehaviour	Theft of produce and damage to land and crops caused by neighbours	Lavergne, 2004, Orsini et al.,2013
Improper recruitment of people with disabilities fitting the initiative	Poor management and inefficient distribution of tasks among personnel with disabilities	*Field outcome*

of labour availability, and undeveloped marketing pathways. Access to land due to urbanisation and insecure land tenure issues were the most highlighted perturbations in the literature, especially in developing countries.

Methods

The methodological approach adopted in the present work uses a *cross-over track analysis*, which emerges as a fundamental criterion to create a mixed-method approach involving both qualitative and quantitative analyses (Tashakkori & Teddlie, 2010). In particular, in such an analysis "findings from the various methodological strands intertwine and inform each other throughout the study" (Datta, 2001, p. 34). According to Teddlie and Tashakkori (2009), this is a more complex form of parallel mixed analysis (Li et al., 2000) characterised by the following: (a) having more than two strands in the design, (b) allowing the analytical strands to inform each other before the meta-inference stage, and (c) consolidating the qualitative and quantitative data such that data are analysed together. Hence, the combination of quantitative and qualitative data helped us study both vulnerability, resilience, and their relation, that way creating a critical reparative approach.

Our work is based on four case studies. Case selection started by compiling a list of potential projects. Using conversations with key stakeholders and an internet search, we identified nine urban agriculture projects in four Moroccan cities: Kénitra, Salé, Casablanca, and Marrakech. Representatives of these nine projects were interviewed using a semi-structured interview guide designed by Knapp (2013) for her research on urban agriculture vulnerability (on which the paper by Knapp et al. (2016) was based). These interviews formed a starting point to better understand each project's main vulnerabilities and to check their availability and willingness to participate in the study. This preliminary investigative survey led to five case-specific perturbations, which we included in the previously developed list distilled from the literature (recall Table 3.1). Four of the nine projects expressed a willingness to participate in the research. They were all included. We shortly introduce the case studies below: their main characteristics are summarised in Table 3.2.

- CIAT is an initiative of the National Centre for Work-Based Integration and Assistance that promotes access to employment for youth with mental disabilities. The pilot project was initiated in 2016 to promote the professional integration and empowerment of young people with mental disabilities in the agricultural sector through on-the-job training on an organically certified farm (market gardening, a nursery, and fruit production). The youths are accompanied by agricultural workers and supervised by specialised educators and technicians.
- Bayti is an organic farm school initiative of Bayti Association, a non-profit NGO for the protection and psycho-social reintegration of children in difficult situations and for the defence of their rights. Since 2005, it has offered an alternative for the integration of teenagers and young adults at risk through a 2-year vocational training programme leading to a diploma in agricultural trades, combined with psychosocial rehabilitation work. This initiative is located on rented land in a peri-urban area and mainly produces strawberries and citrus fruits. Two years after our research, it acquired the high-value environmental certification from the French association "Demain la Terre."
- Dar Bouazza is an individual, self-financed project that started in 2017 based on the idea and technical support provided by Terre et Humanisme Maroc (THM), a pioneer association of agroecology-based solidarity in Morocco. Two years after our research, this vegetable and animal farm managed by an active farmer on

Table 3.2 Characteristics of the selected case studies

Project	*CIAT*	*Bayti*	*Dar Bouazza*	*UAC-PP4*
City	Periphery of Salé	Periphery of Kenitra	Periphery of Casablanca	Periphery of Casablanca
Year of creation	2016	2005	2017	2005–2013
Surface	7 ha	13 ha	3 ha	0.8 ha
Initiator(s)	The national centre Mohamed VI for persons with disabilities; Foundation Mohamed V for solidarity	Association Bayti	Terre et Humanisme association	Producer; UAC project
Partners	Crédit agricole; OFPPT	UNICEF, Drosos, GIZ & the Spanish agency for international cooperation	Landowner	29 Moroccan German partners
Form	Social entrepreneurship initiative	Social entrepreneurship initiative	Family commercial garden	Social entrepreneurship initiative
Function	Pedagogic; social inclusion	Social inclusion	Income generation	Pedagogic
Organic certification	Yes	Yes	No	Yes
Marketing	On farm sale; restaurant	Local market; exports to USA and France	Box delivery; on farm sale; sale to a retailer: Carrefour	Box delivery; on farm sale
Interviewees	5	4	5	3
Role	Manager, technician, assistant, nurseryman, and quality manager	Director, technician, workers, and educator	Farmer, son, client, neighbour, and sale manager of Carrefour	Landowner, project collaborator, and coordinator

a rented peri-urban plot has doubled its area of production, hired new workers, and created new steady market linkages.

- UAC-PP4 is a pilot initiative, Pilot Project 4 (PP4), on healthy food production and urban agriculture implemented in the framework of the Moroccan-German project "Urban Agriculture in Casablanca" (UAC). It combines urban and organic agriculture, aiming at awareness raising of urban consumers about the importance of local food consumption. Two years after our research, it switched its cropping system from horticultural crops to fruit trees to overcome climate change effects. It also started organising awareness events and hosting team building activities as part of corporate social responsibility and for revenue integration.

A first round of data collection and analysis (summer 2018) was both quantitative and qualitative in nature, with the aim to obtain a composite vulnerability score. Stakeholders from each of the four selected projects were interviewed using semi-structured interview guides, consisting of both qualitative and quantitative questions regarding the perturbations listed in Table 3.1. For each perturbation, we first used an open question, asking about the extent to which the perturbation was considered relevant, after which we scrutinised the perturbation in more detail, based on the vulnerability framework. We discussed exposure, sensitivity, and resilience by operationalising these concepts into questions in more colloquial language. With respect to exposure, for instance, we asked how often a perturbation was encountered, for how long, and how severe the consequences of that specific perturbation were. Besides these qualitative conversations, we asked interviewees to score resilience, exposure, and sensitivity on a five-point Likert scale (1 = very low, 5 = very high); the resilience scale was inverted as it is considered a strength, whereas exposure and sensitivity are weaknesses. Finally, interviewees were asked to score the importance of each perturbation. Respondents were selected to represent different views on the case studies, such as consumers, workers, managers, farmers, and directors. We interviewed three to five respondents per case (see Table 3.2). All interviews were recorded and transcribed ad verbatim.

In summer 2020, we conducted follow-up interviews with a representative of each case study to gain an update regarding the perturbations and the way in which the cases dealt with them. These interviews were performed over the phone.

We used NVivo software to code interview transcripts, based on a predetermined set of codes, containing the list of perturbations previously determined. For each of the four case studies, we then calculated a vulnerability score for all 24 perturbations, in the following way (also see Table 3.3):

- We scored the levels of exposure, sensitivity and resilience for each perturbation, using the Likert scale as filled out by respondents and calculating mean values when more than one interviewee of a case scored these elements (QUAN in Table 3.3).
- The interviewer also provided personal scores based on the information collected in interviews (coded with NVivo) (QUAL in Table 3.3). Adding our own score based on the qualitative data helped us make better comparisons between cases.
- These two scores (QUAL and QUAN) were then used to create intermediate scores for exposure, sensitivity, and resilience.
- Vulnerability scores, finally, were obtained as the average of the intermediate scores of exposure and sensitivity, and the reversed score of resilience.

Table 3.3 Data analysis scheme

<table>
<tr><th>Perturbation</th><th colspan="3">Exposure</th><th colspan="3">Sensitivity</th><th colspan="3">Resilience</th><th>Vulnerability Scores</th></tr>
<tr><th></th><th>QUAN</th><th>QUAL</th><th>Exp. (score)</th><th>QUAN</th><th>QUAL</th><th>Sen. (score)</th><th>QUAN</th><th>QUAL</th><th>Res. (score)</th><th></th></tr>
<tr><td>Perturbation 1</td><td>mean value</td><td>NVivo coding</td><td>Inter. Score</td><td>mean value</td><td>NVivo coding</td><td>Inter. score</td><td>mean value</td><td>NVivo coding</td><td>Inter. score</td><td>Mean Exp.|Sen.|Res. rev.</td></tr>
<tr><td>Perturbation 2</td><td>mean value</td><td>NVivo coding</td><td>Inter. Score</td><td>mean value</td><td>NVivo coding</td><td>Inter. score</td><td>mean value</td><td>NVivo coding</td><td>Inter. score</td><td>Mean Exp.|Sen.|Res. Rev.</td></tr>
<tr><td>Perturbation n</td><td>mean value</td><td>NVivo coding</td><td>Inter. Score</td><td>mean value</td><td>NVivo coding</td><td>Inter. score</td><td>mean value</td><td>NVivo coding</td><td>Inter. score</td><td>Mean Exp.|Sen.|Res. Rev.</td></tr>
</table>

Vulnerability and resilience of four Moroccan case studies

Figures 3.2a–3.2d show the composite vulnerability scores of the four case studies. For each recorded perturbation, the figures record the intermediate scores of exposure and sensitivity, and the reversed score of resilience. Taken together, the vulnerability score tells us to what extent each case study is vulnerable towards perturbations in terms of exposure, sensitivity, and resilience. Table 3.4 presents the cross-case vulnerability scores by perturbation. As not all perturbations found in the literature affect our case studies – land contamination, dispensation law, and absence of policies were not recognised as perturbations by any of the cases – the analyses shown in Figure 3.2 and Table 3.4 contain 21 rather than 24 perturbations.

Table 3.4 shows that Dar Bouazza has the highest vulnerability score (4.1). It also has the highest number of recorded perturbations (14). The initiative is particularly exposed and sensitive, and the reversed resilience scores are low for most perturbations due to the limited flexibility of the mechanisms deployed to deal with these perturbations. Bayti is the least vulnerable case, with a vulnerability score of 2.5. Even though several perturbations score high, they do not affect the project to a large extent, as sensitivity is low and the high scores of resilience mitigate the project's overall vulnerability. In other words, the case faces troubles but finds ways to deal with them. CIAT and UAC-PP4 show medium vulnerability scores of 3.6 and 3.5, respectively. CIAT suffers mainly from two perturbations: administrative constraints and undeveloped market pathways. For both of them, exposure and sensitivity is high, and resilience is low and insufficient to mitigate them. UAC-PP4 is exposed and sensitive to two main agricultural perturbations – bad quality of irrigation and climatic hazards – to which the project has not developed adequate defence mechanisms.

Below we discuss eight perturbations in more detail. We chose to feature these eight, which cover all five perturbation categories (see Table 3.1), based on the following criteria: (1) their substantive presence within the analysed projects, and (2) their high impact in terms of vulnerability score. Hence, focusing on these perturbations enables us to show the main troubles these initiatives face and the ways in which they manage them. Table 3.4 highlights the selected perturbations in bold font and grey cells and shows their corresponding vulnerability scores per case study as well as the occurrence and the total vulnerability score of each.

Project location and structural features

Insecure land tenure

Insecure land tenure showed to be a significant difficulty for three of the studied projects, although they all perceived the perturbation differently. Contrary to what Knapp et al. (2016) reported, all projects were clear about the reasons behind their concerns towards this perturbation. Bayti project does not consider land tenure as particularly risky as the land used belongs to the king and its lease is renewed every ten years. For Dar Baouzza, on the other hand, land tenure is very insecure

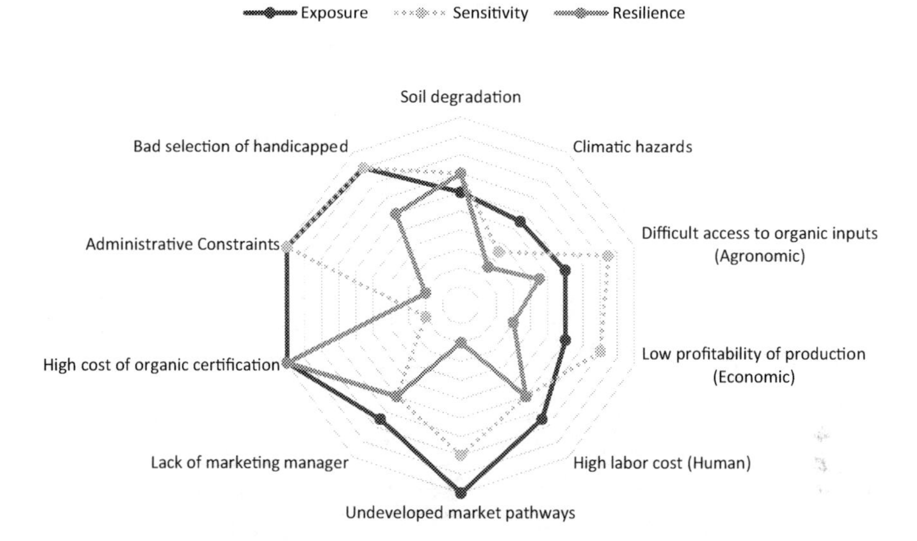

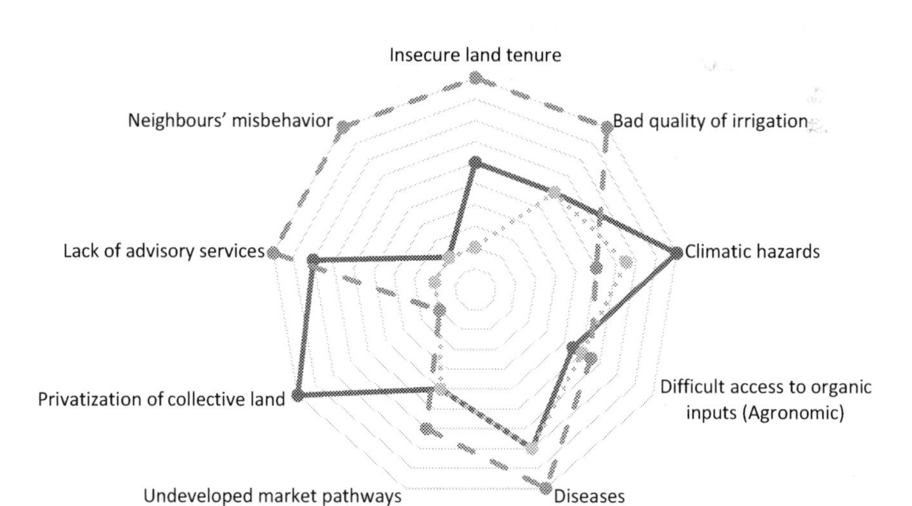

Figure 3.2 Composite vulnerability score of the four case studies

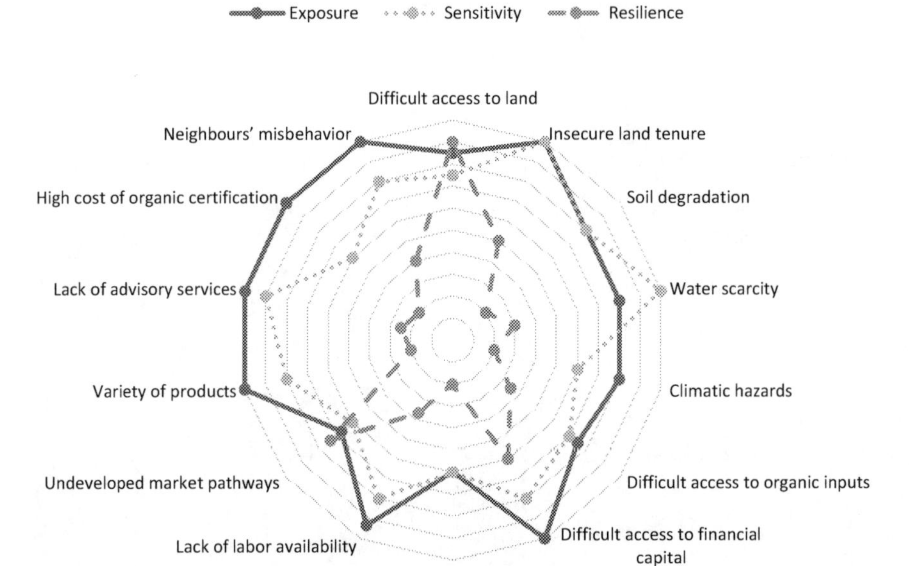

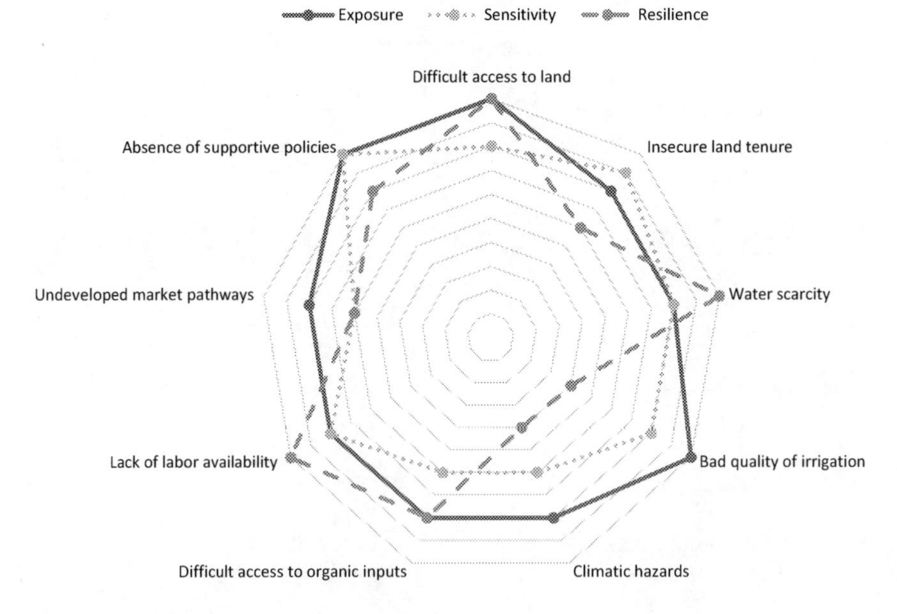

Figure 3.2 Continued

Table 3.4 Cross-case vulnerability scores by perturbation

Perturbations		*Vulnerability scores*				*Case occurrence*	*Total vulnerability score*
		CIAT	*Bayti*	*Dar Bouazza*	UAC-PP4		
1. Project location and structural features	**Difficult access to land**			3.2	3.3	2	6.5
	Insecure land tenure		1.7	4.5	3.8	3	10
	Soil degradation	3		4.3		2	7.3
	Water scarcity			4.5	3	2	7.5
	Bad quality of irrigation water		2.3		4.3	2	6.6
	Climatic hazards	3.1	3.9	4	3.7	4	14.7
2. Project management	**Difficult access to organic inputs**	3.7	2.8	3.8	3	4	13.3
	Difficult access to financial capital			4		1	4
	Low profitability of production	3.8		3.7		2	7.5
	Lack of labour availability and qualification	3.2		4.3	3	3	10.5
	Diseases		3			1	3
3. Interaction with market actors	**Undeveloped market pathways**	4.7	2.5	2.9	3.3	4	13.4
	Lack of marketing manager	3.3				1	3.3
	Lack of variety of products			4.7		1	4.7
4. Interaction with institutional settings	Absence of urban supportive policies				4	1	4
	Privatisation of collective land		3.7			1	3.7
	Lack of advisory services		2	4.8		2	6.8
	High cost of organic certification	2.3		4.3		2	6.6
	Administrative constraints	5				1	5
5. Interaction with society	**Neighbours' misbehaviour**		1	4.3		2	5.3
	Improper recruitment of people with disabilities fitting the initiative	4				1	4
Vulnerability score		3.6	2.5	4.1	3.5		

as the project has an informal contract for its use. The land can be given to others, which has already happened in the past. This perturbation is aggravated as the farm is organic: the transition required a considerable amount of time. The farmer is therefore reluctant to plant trees, even if this is evaluated as a good opportunity. Such a preclusion to invest in infrastructure and the limited choice of cultivated species in order to keep cropping cycles short were reported by Lorenz (2015) and Orsini et al. (2013), respectively. For UAC-PP4 land tenure is a cause of vulnerability, as the use of land is guaranteed only for the next ten years. It is currently included in a green zone of the city layout plan, but it is unclear what will happen after this term.

Climatic hazards

Climatic hazards were reported as a source of vulnerability in literature until the 1990s (Callo-Concha & Ewert, 2014). They were detected as troublesome in all four projects and showed to be mainly significant in terms of exposure. Climatic hazards are directly related to the geoclimatic conditions of the case study locations: the projects are sited in a coastal area with frequent high-intensity winds and alternated periods of drought and flood and lately prolongated high temperatures. These weather conditions damage crops and/or equipment, generating additional costs to repair and implement adaptive, resilient solutions. (Please note that we did not discuss climate change as such but approached this perturbation as related to the geographic location of the cases.) Both Bayti and UAC-PP4 implemented actions to mitigate this climate risk: Bayti chose to increase resilience by planting hedges of trees, UAC-PP4 created water basins, and applies specific organic techniques like crop rotation, temporary fallow plots, and the use of adapted local varieties. Unfortunately, these solutions are not easily applicable to CIAT and Dar Bouazza. For CIAT, the lack of managerial decisions slows down targeted interventions to contrast this perturbation, while for Dar Bouazza, land tenure insecurity inhibits such interventions, thus resulting in a very low resilience score.

Project management

Difficult access to organic inputs

When organically certified, farmers can only use authorised materials such as fertilisers, soil conditioners, substrates, plant protection products, cleaning agents, and disinfectants. Such organic inputs are considered expensive and limitedly accessible; this affects harvests' quantity, quality, and variety, and consequently, quantity and quality of marketable produce. However, although all case studies are organic, this type of trouble influences them to different degrees. Bayti and UAC-PP4 projects were able to develop a high resilience towards the difficult access to organic inputs by stocking large quantities when they became available on the market. Lorenz (2015) reported that one of the potential benefits of organic agriculture is the resulting social network that can provide technical or financial support. However, none of our four case studies referred to or mentioned having tried to get input through their networks or supportive institutions.

Lack of labour availability and qualification

Three of the case studies were found to be vulnerable to a lack of available skilled wage labourers, with high exposure scores. However, it is considered an important source of vulnerability mainly for Dar Bouazza, as this case sports high sensitivity and low resilience. Previous studies confirmed the difficulty of accessing manpower (De Bon et al., 2008; Orsini et al., 2013) and attracting young people to agriculture. This is mainly due to the perceived low status of the job and the limited scale of mechanisation, making it hard work. Moreover, potential workers prefer big industries to agriculture, as working in big industries guarantees higher salaries and long-term contracts, which our case studies cannot offer. Some cases hope that immigrant workers will solve the problem, but the long paperwork process and a general distrust of foreigners discourage potential workers. Especially UAC-PP4 struggles with this perturbation due to strong competition on the labour market, but it developed a high resilience by sharing responsibility with existing workers, that way creating motivational ownership spirits.

Interactions with market actors

Undeveloped marketing pathways

Underdeveloped marketing pathways were found to be a perturbation in all the case studies, with high exposure and sensitivity but different resilience reactions. The four cases market their products mainly through direct sales and box deliveries, both short supply chains commonly used by urban agriculture initiatives in developing countries (Moustier & Renting, 2015). CIAT reduced its prices to tackle low consumer buying capacity and limited awareness about organic products. As this proved to be an inefficient measure to resolve the marketing problem as such (i.e. reaching consumers), the project remains highly vulnerable – the trouble is not solved. Bayti reacted by enlarging production capacity and engaging in export activities. This diminished exposure to the perturbation, making the project more resilient. Dar Bouazza faces serious problems with product assortments and the organisation of delivery, leading to consumer complaints and thus high exposure and sensibility. The project owner attributes the project's high sensibility to this perturbation to his illiteracy and unfamiliarity with digital means of communication, which limit him from using social media and technology to reach aware and rich consumers. Nevertheless, its resilience is high due to its sustainable marketing channel through Carrefour, a French supermarket chain well-known in Morocco. For UPC-PP4, the lack of an organised market and its consumers being unsatisfied led many workers to abandon the project. This increased not only the exposure and sensitivity to this perturbation but also the trouble around the lack of labour availability.

Interactions with institutional settings

Lack of advisory services

The lack of advisory services was strongly pinpointed as a source of vulnerability in two of the projects. Although both Bayti and Dar Bouazza registered high exposure scores to this perturbation, the projects reacted differently. Bayti took advantage of external technical help and easily overcame knowledge gaps by relying on the networks of its supporting partners for exchanging and sharing knowledge. Dar Bouazza sought advice from its limited relationships with input suppliers and neighbouring organic farmers, as also reported by Giseke et al. (2015). However, when such support exists, it is often run by associative entities that do not usually accompany projects permanently, not completely solving the trouble. This solution was therefore not enough to tackle the problem and raise resilience in relation to this perturbation.

High cost of organic certification

The high cost of organic certification was recorded as a vulnerability in two projects only. It affects Dar Bouazza in an extreme fashion, with high exposure and sensitivity and very low resilience. As with many urban farms (Lorenz, 2015), Dar Bouazza couldn't get its organic certification due to the very complex and costly procedures. While highly exposed, CIAT showed low sensitivity and high resilience towards the high costs of organic certification, as these were mitigated by its outputs and the charity-sponsored activities of the project. Other modes of certification, or establishing personal trust relations with buyers (e.g. participatory certification), did not come up in interviews.

Interactions with society

Neighbours' misbehaviour

This perturbation was registered only in Bayti and Dar Bouazza projects. However, while Bayti had a low vulnerability score – both exposure and sensitivity were low and resilience was high – Dar Bouazza had high vulnerability as exposure and sensitivity were high and resilience low. Resilience to this perturbation is high for Bayti as neighbourhood youngsters, who were considered a potential source of disturbance, were included in project trainings so as to raise their awareness of the importance of the initiative and build a relationship of trust and respect. This supports the finding of Robineau et al. (2014), who stated that the older social relations with the neighbourhood are, the more tolerance they show, and the smaller the risk of complaints to the authorities. Resilience was low for Dar Bouazza as the landowner's family members take advantage of their family bond by frequently harvesting vegetables and other fresh produce without permission. This leads to an insufficient harvest for the box schemes. The absence of sharp and clear rules for the use of the project's production leads to a complicated situation that causes a high vulnerability towards this perturbation.

Discussion and conclusions

All four cases presented in this chapter deal with a number of perturbations, or in the words of this book: troubles. While these troubles show considerable overlap, the degree to which the four cases are able to deal with them differs. As a result, both vulnerability and resilience vary largely. Hence, whereas cases may struggle with similar difficulties, some of them are better able to turn these into more hopeful situations, and similar types of troubles are experienced differently. Reacting to or dealing with perturbations can thus be a source of change, as finding solutions for these perturbations means that different directions are taken, resources are used differently, or strategies are adjusted. In some cases, this indeed creates hope. In other cases, however, it may lead to new perturbations or increased vulnerability towards others.

Despite these difficulties, the fact that the cases show a certain degree of resilience to most perturbations faced, is also an element of hope. Here too, however, the differences between cases are large. In general, we found that older projects such as Bayti tend to be more resilient than younger projects like Dar Bouazza. Apart from that, the sources of resilience are multiple and varied, but often relate to social-institutional characteristics: the establishment of a good relationship with neighbours, a broad network to create access to financial or technical support, and stock management to secure organic inputs help tackle the perturbations faced. These sources of hope are unequally distributed between projects, however. This is illustrated by the example of different profiles of project leaders: whereas the project leader of Dar Bouazza is an illiterate farmer unable to use social media or smart technologies to promote his activity, the project leader of UAC:PP4 has an academic background, speaks different languages, and has a presidential position in an association. This helped the project find land and attract consumers. Dar Bouazza, on the other hand, is barely surviving. The project leader lacks the resources, time, and energy that could be used to work on certain perturbations to reduce his overall vulnerability. This finding confirms those of other chapters in this volume: social inequality is indeed a major problem in urban agriculture and community food initiatives more broadly.

The vulnerability framework helped us draw out both the troubles projects face (perturbations), and their hopes (their sources of resilience), illuminating that even when difficulties are present, initiatives may fight and overcome them. It also shows that certain situations or events may both hamper and strengthen resilience, depending on local situations and specificities. Organic farming serves as an example. All projects studied work organically, which proved to have a varying impact. In some cases, organic farming increased vulnerability as organic inputs, technical knowledge, and marketing skills were difficult to access: it was a source of trouble. In other cases, however, organic farming was a source of hope, having added value in terms of marketing and network support. For instance, farming organically alleviates Bayti's vulnerability due to its close work with international sponsors and partners. Working on both organic agriculture and the social aspects of housing and educating homeless children gives the project a strong push to be recognised internationally and to compete nationally. In the words of the project leader: *Organic agriculture connects you with a specific intellectual niche so that you can select people with whom to work*. Clearly, perturbations are not equally troublesome.

To conclude, the cases presented in this chapter exemplify a plurality of hopes and troubles in their composite, multi-level struggle for resilience. Taken together, the snapshot of the four cases represents a kaleidoscopic picture of these hopes and troubles that is highly case specific. Our work therefore corroborates the main premise of this book, confirming that it is essential to go beyond either an "all good" narrative or an essentialist critical stance, and to look at the hybrid complexity and fluidity of initiatives. In order to produce effective, more permanent change on the ground in local food systems, an inevitable and crucial step for the near future is to raise awareness of the many points of contacts and complementarities between such initiatives, and of the need for various stakeholders – academics, policy makers, NGOs – to join forces, approaches and critical thinking. We also urge other researchers to work at a more comprehensive and inclusive design of mixed analysis, where both quantitative and qualitative data occur concurrently (not in chronological order), sequentially in two phases (the qualitative analysis phase precedes the quantitative phases or vice versa), or even envisage more than two phases (iteratively) (Tashakkori & Teddlie 2010).

References

Adger, W. N. (2006). Vulnerability. *Global Environmental Change*, *16*(3), 268–281.

Aubry, C., & Pourias, J. (2012). L'agriculture urbaine fait déjà partie du "métabolisme urbain." In *Déméter 2013: Club déméter* (p. 432). https://hal.archives-ouvertes.fr/hal-01198075

Banzo, M., Perrin, C., Soulard, C.-T., Valette, E., & Mousselin, G. (2016). Rôle des acteurs publics dans l'émergence de stratégies agricoles des villes, exemples en Méditerranée. *Economia e Società Regionale*, *23*, 8–30.

Barbizan, T. S. (2011). *Integrating urban and peri-urban agriculture into public policies to improve urban growth: São Paulo as a case study* (p. 166). University of Technology.

Benabed, A., Dugué, P., & Abdellaoui, E. H. (2014). Les exploitations familiales peuvent-elles faire face à l'urbanisation? Cas de la commune urbaine de Sebaa-Ayoune dans la plaine du Saïs (Maroc). *Alternatives Rurales*, *1*, 11.

Callo-Concha, D., & Ewert, F. (2014). Using the concepts of resilience, vulnerability and adaptability for the assessment and analysis of agricultural systems. *Change and Adaptation in Socio-Ecological Systems*, *1*, 1–11. http://dx.doi.org/10.2478/cass-2014-0001

Datta, L.-E. (2001). The wheelbarrow, the mosaic and the double helix: Challenges and strategies for successfully carrying out mixed methods evaluation. *Evaluation Journal of Australasia*, *1*(2), 33–40. https://doi.org/10.1177/1035719X0100100210

Davidson, J. D. (2017). Is urban agriculture a game changer or window dressing? A critical analysis of its potential to disrupt conventional agri-food systems. *International Journal of Sociology of Agriculture and Food*, *23*(2), 63–76.

Debolini, M., Valette, E., François, M., & Chéry, J.-P. (2015). Mapping land use competition in the rural – urban fringe and future perspectives on land policies: A case study of Meknès (Morocco). *Land Use Policy*, *47*, 373–381. http://dx.doi.org/10.1016/j.landusepol.2015.01.035

De Bon, H., Parrot, L., & Moustier, P. (2010). Sustainable urban agriculture in developing countries: A review. *Agronomy for Sustainable Development*, *30*(1), 21–32. http://dx.doi.org/10.1051/agro:2008062

De Zeeuw, H., Van Veenhuizen, R., & Dubbeling, M. (2011). The role of urban agriculture in building resilient cities in developing countries. *The Journal of Agricultural Science*, *149*(S1), 153–163. http://dx.doi.org/10.1017/s0021859610001279

Dugué, P., Benabed, A., Abdellaoui, E. H., & Valette, E. (2015a). L'agriculture urbaine à Meknès (Maroc) à la croisée des chemins: Disparition d'une agriculture marginalisée ou retour de la cité jardin? *Alternatives Rurales*, *3*, 15.

Dugué, P., Soulard, C.-T., Marraccini, E., Houdart, M., Michel, I., & Rhaidour, M. (2016). Systèmes maraîchers urbains et périurbains en Méditerranée: Une comparaison entre Meknès (Maroc), Montpelier (France) et Pise (Italie). In H. Rejeb & C.-T. Soulard (Eds.), *Organisation des agriculteurs et des systèmes agricoles dans les territoires urbains et périurbains* (pp. 57–78). Tunisia Institut français de Tunis, ISA-IRESA, INRA, CIRAD.

Dugué, P., & Valette, É. (2015). Des agriculteurs marginalisés au cœur des villes: Le cas de Meknès (Maroc). *Pour*, *225*, 61–67. https://doi.org/10.3917/pour.225.0061

Dugué, P., Valette, E., Chia, E., Abdellaoui, E. H., & Vitry, C. (2015b). Le concept de système agri-urbain confronté aux réalités des villes du sud de la Méditerranée: Le cas de Meknès (Maroc). In *52ème colloque: Territoires Méditerranéens: Agriculture, alimentation et villes* (pp. 1–16). Association de sciences régionales de langue française.

Eakin, H. C., & Wehbe, M. B. (2009). Linking local vulnerability to system sustainability in a resilience framework: Two cases from Latin America. *Climatic Change*, *93*(3–4), 355–377.

Giseke, U., Gerster-Bentaya, M., Helten, F., Kraume, M., Scherer, D., Spars, G., Adidi, A., Amraoui, F., Berdouz, S., Chlaida, M., Mansour, M., & Mdafai, M. (2015). *Urban agriculture for growing city regions: Connecting urban-rural spheres in Casablanca*. Routledge.

Jouve, A. M., & Padilla, M. (2007). Les agricultures périurbaines méditerranéennes à l'épreuve de la multifonctionnalité: Comment fournir aux villes une nourriture et des paysages de qualité? *Cahiers Agricultures*, *16*(4), 7. http://dx.doi.org/10.1684/agr.2007.0109

Knapp, L. (2013). *Implementing urban agriculture in Europe: A case study of urban agriculture projects in the Netherlands and Switzerland* [Master thesis, Wageningen University, Ecological Farming Systems].

Knapp, L., Veen, E., Renting, H., Wiskerke, J. S. C., & Groot, J. C. J. (2016). Vulnerability analysis of urban agriculture projects: A case study of community and entrepreneurial gardens in The Netherlands and Switzerland. *Urban Agriculture and Regional Food Systems*, *1*, 1–13. http://doi: 10.2134/urbanag2015.01.1410

Lavergne, M. (2004). L'agriculture urbaine dans le bassin méditerranéen, une réalité ancienne à l'heure du renouveau. In J. Nasr & M. Padilla (Eds.), *Interfaces: Agricultures et villes à l'Est et au Sud de la Méditerranée* (pp. 49–66). Éditions Delta/IFPO.

Li, S., Marquart, J. M., & Zercher, C. (2000). Conceptual issues and analytic strategies in mixed-method studies of preschool inclusion. *Journal of Early Intervention*, *23*(2), 116–132. https://doi.org/10.1177/105381510002300206

Lorenz, K. (2015). Organic urban agriculture. *Soil Science*, *180*(4–5), 146–153. http://dx.doi.org/10.1097/ss.0000000000000129

Marsden, T., Hebinck, P., & Mathijs, E. (2018). Re-building food systems: Embedding assemblages, infrastructures and reflexive governance for food systems transformations in Europe. *Food Security*, *10*, 1301–1309.

Moustier, P. (2017). Short urban food chains in developing countries: Signs of the past or of the future? *Natures Sciences Sociétés*, *25*(1), 7–20.

Moustier, P., & Renting, H. (2015, January). Urban agriculture and short chain food marketing in developing countries. In H. de Zeeuw & P. Drechsel (Eds.), *Cities and agriculture – developing resilient urban food systems* (pp. 121–138). Earthscan Food and Agriculture. https://doi.org/10.4324/9781315716312

Orsini, F., Kahane, R., Nono-Womdim, R., & Gianquinto, G. (2013). Urban agriculture in the developing world: A review. *Agronomy for Sustainable Development*, *33*(4), 695–720. http://dx.doi.org/10.1007/s13593-013-0143-z

Robineau, O., Tichit, J., & Maillard, T. (2014). S'intégrer pour se pérenniser: Pratiques d'agriculteurs urbains dans trois villes du Sud. *Espaces et Sociétés*, *158*(3), 83. http://dx.doi.org/10.3917/esp.158.0083

Rosol, M. (2020). On the significance of alternative economic practices: Reconceptualizing alterity in alternative food networks. *Economic Geography*, 96(1), 52–76. https://doi.org/10.1080/00130095.2019.1701430

Rousseau, M., Boyet, A., & Harroud, T. (2020). Politicizing African urban food systems: The contradiction of food governance in Rabat and Casablanca, Morocco. *Cities*, *97*, 102528. https://doi.org/10.1016/j.cities.2019.102528

Sarmiento, E. R. (2017). Synergies in alternative food network research: Embodiment, diverse economies, and more-than-human food geographies. *Agriculture and Human Values*, *34*(2), 485–497. https://doi.org/10.1007/s10460-016-9753-9

Soussan, C. (2015). *Reconnaissance de l'agiculture urbaine et périurbaine dans les métropoles méditerranéennes: Etudes de cas* (p. 105). Sciences de l'homme et société.

Sovová, L., & Veen, E. J. (2020). Neither poor nor cool: Practising food self-provisioning in allotment gardens in the Netherlands and Czechia. *Sustainability*, *12*(12), 5134. https://doi.org/10.3390/su12125134

Tashakkori, A., & Teddlie, C. (2010). *Sage handbook of mixed methods in social & behavioral research* (2nd ed.). Sage Publications. https://dx.doi.org/10.4135/9781506335193

Teddlie, C., & Tashakkori, A. (2009). *Foundations of mixed methods research: Integrating quantitative and qualitative approaches in the social and behavioral sciences*. Sage Publications.

Turner, B. L., Kasperson, R. E., Matson, P. A., McCarthy, J. J., Corell, R. W., Christensen, L., Eckley, N., Kasperson, J. X., Luers, A., Martello, M. L., Polsky, C., Pulsipher, A., & Schiller, A. (2003). A framework for vulnerability analysis in sustainability science. *Proceedings of the National Academy of Sciences of the United States of America*, *100*(14), 8074–8079. http://dx.doi.org/10.1073/pnas.1231335100

Urruty, N., Tailliez-Lefebvre, D., & Huyghe, C. (2016). Stability, robustness, vulnerability and resilience of agricultural systems: A review. *Agronomy for Sustainable Development*, *36*(1), 15. http://dx.doi.org/10.1007/s13593-015-0347-5

Valette, E., & Dugué, P. (2017). L'urbanisation, facteur de développement ou d'exclusion de l'agriculture familiale en périphérie des villes: Le cas de la ville de Meknès, Maroc. *VertigO – La Revue Électronique en Sciences de L'environnement*, *17*(1), 24. http://dx.doi.org/10.4000/vertigo.18413

Valette, E., & Philifert, P. (2014). L'agriculture urbaine: Un impensé des politiques publiques marocaines?, *Géocarrefour*, *89*(1–2), 75–83.

Valette, E., Schreiber, K., Conaré, D., Bonomelli, V., Blay-Palmer, A., Bricas, N., Sautier, D., & Lepiller, O. (2020). An emerging user-led participatory methodology: Mapping impact pathways of urban food system sustainability innovations. In A. Blay-Palmer, D. Conaré, K. Meter, A. Di Battista, & C. Johnson (Eds.), *Sustainable food system assessment: Lessons from global practice*. Routledge Studies in Food, Society and Environment.

Zasada, I. (2011). Multifunctional peri-urban agriculture – a review of societal demands and the provision of goods and services by farming. *Land Use Policy*, *28*(4), 639–648. http://dx.doi.org/10.1016/j.landusepol.2011.01.008

4 Spaces of hope and realities beyond the fence

Experiences of urban food providers in South Africa

Anne Siebert

Introduction

Mounting concerns about the negative impacts of the industrialised agri-food system and several interconnected global crises – in climate, human health, displacement, and democracy, just to name a few – fuel the work of community food initiatives (CFI). Certainly, the proliferation of CFIs is intertwined with the inequality-enhancing machine of capitalism including its disruptive consequences for the environment and human well-being. Eric Olin Wright provides us with a profound critique of capitalism. He actively engages with society and illuminates a wide range of issues like precarious livelihoods and inappropriate working conditions. But more importantly, he is introducing strategies of *hope*; he is convinced that "another world is possible" (2019, p. 3). These fresh analytical perspectives are much needed to understand and fuel the transformative efforts of CFIs.

Like in many other countries, urban CFIs in South Africa intend to create alternatives to exclusive food and agriculture landscapes, which speaks to the notion of *hope* and reparative intentions (cf. Siebert, 2020; Tornaghi, 2017). For instance, they oppose prevailing power structures, which make it difficult for marginalised population groups to access and produce healthy food. South African cities and the agri-food system tell an intense story of racialised inequality. With approximately two-thirds of the population living in cities in 2018, 27.1% of the population was unemployed, and striking malnutrition[1] rates suggest that about one-fifth of urban households experience hunger (United Nations Population Division, 2018; Statistics South Africa, 2018a). These numbers are higher for black South Africans. The country's strong neoliberal turn is in favour of large-scale agriculture and dominant supermarket chains, which pose diverse challenges to small-scale farmers and marginalised consumers. Against this background, urban agriculture comprises diverse functions, for instance in self-provisioning, social cohesion, or healthier diets; these functions may also comprise political activism, notions of protest, and also innovative ways of overcoming restrictions in food and income (Siebert, 2020).

In sum, the outlined inequalities are the breeding ground for "responses from below," reaching from protests to simplified self-help mechanisms in urban gardening. However, the deep roots and manifestations of poverty and inequality seem to limit and challenge broader changes. Responses, including CFIs, are often localised

DOI: 10.4324/9781003195085-5

and isolated – mirroring the notion of *trouble* discussed in this volume. These issues are also evident in the Erf[2] 81 community in Cape Town, South Africa – a community farm and kind of a solidarity network, which is the focus of this contribution. One of the pending questions in this context is raised by Chiara Tornaghi,

> under what conditions can urban agriculture escape its marginality and contribute to reimagining, reshaping and radically changing the food system, and in doing so liberate us – at least partially – from the absolute capitalist control over a fundamental sphere of social reproduction?
>
> (2017, p. 791)

This question confronts us with the key challenges of CFIs and their almost impossible demands. Still, such questions are key to advancing and improving CFIs noble work. Using this inspiration and following a critical reparative approach, the guiding question of this contribution is, how can alternative food providers truly create hope through resisting the system in place and confront related troubles?

Using Wright's work on "How to be an anticapitalist in the 21st century" (2019) as part of the critical reparative approach, it is the attempt of this chapter to provide more insights into CFIs' promising actions, but particularly restricting factors and ways to overcome these. In this way, the key aim is to carve out possibilities beyond difficulties.

This chapter is structured in the following way. First, I briefly introduce the Erf 81 community. In the second part, I sketch out some key features of South Africa's unequal agri-food system, which provide a starting point for Wright's typology of anti-capitalist strategies and the related analytical lens. Specifically, this analytical lens is presented. In the main part, I use Wright's strategies to illuminate the transformative potential of the Erf 81 initiative and its limitations. The combination of the theoretical notions and the lived realities on the ground are essential to carve out *troubles* among often more visible *hopes*. I moreover show possibilities to create a broader impact beyond a localised sphere. The last part provides a short conclusion. The outlined findings are considered helpful for CFIs, policy makers, and activists to critically evaluate and improve revolutionary endeavours.

Erf 81 community and urban agriculture

The following insights on the Erf 81 community are based on a qualitative and explorative research approach comprising personal visits to the farm, informal exchange with members (in both 2016 and 2019), insights provided in the media (e.g. newspaper articles, YouTube clips), and a few community-related academic publications. I analysed this material with Wright's typology of "anti-capitalist strategies" to contrast the initiative's transformative efforts against the exclusionary agri-food system and social system (2017, 2019). This part presents a first introduction to the CFI.

Some call the Erf 81 community "a farm in the rainbow" (Petrie, 2016), which speaks to both its unique location and its open-minded community. On abandoned military land at the fringes of the central business district and at the slopes of Signal Hill a diverse community of about 30 people grows veggies and herbs and raises livestock on occupied land. The mixed group of black, coloured, and white South Africans as well as a few foreigners builds a harsh contrast to the bordering predominantly affluent white neighbourhood Tamboorskloof and the increasingly gentrified former Malay slave quarter Bo-Kaap. Hence, they vitally break down post-Apartheid barriers, which can still be observed in the city. The group consists mainly of young people between 20 and 40 as well as a few early retirees. Most of them are not fully integrated in the labour market, try to combine different income strategies, and therefore struggle to a certain extent to make a living. They are students and NGO workers, young start-up entrepreneurs, or part-time farmers. The farm basically offers them a home – a place to stay, some food on the table, a family, and for some, an opportunity to make a bit of income through market and cultural activities. Still, the choice for such rather precarious, illegal living conditions is originating to a certain extent in broader social inequalities in South Africa, which, for instance, complicate sufficient access to farm land, education, the employment market, and affordable food and housing. I will come back to this in the next part of this contribution.

The first land occupations and food growing in this location started about 25 years ago, when a "foster farm" for street children was initiated by a young family. It basically offered an alternative place to live for these neglected youngsters. André, who founded the farm and probably is the oldest community member, still seems to be kind of a natural leader. While the children are grown up by now, most of them are still part of the community. Today, the original intention of offering an alternative, creative, sustainable, and peaceful place and way of living can still be felt. In sum, shared aims of solidarity, feeding, reconciling, and protecting the community, vicinity, and nature, as well as the struggle for food sovereignty[3] broadly drive their actions. Food sovereignty can be understood as a powerful – progressive and in part radical – social movement response to the globalised industrialised agri-food system and neoliberal global food politics (Alonso-Fradejas et al., 2015; Siebert, 2020). These values are also broadly mirrored in a quote from one member "agriculture is not for people to get richer" (Petrie, 2016), which shows that profit for a few, like private companies, is not the intention.

Beyond honourable intentions, one has to admit that the farm and its community may appear somewhat chaotic and rebellious to outsiders. The community resides there illegally. For instance, pigs, sheep, and chickens roam around freely, fences do not necessarily exist, and gates are mostly open for visitors. Moreover, there is no clear organisational structure. Still, this commune presents a certain revolutionary potential within existing structures. Gee (2017, p. 22) describes the farm in the following way: "The farm has a feeling of revolt or rebellion against normal societal structure, with many of its barriers being brought down, with the goal of focusing on love, openness and creativity." Existing barriers in this context might be referred to as gender, race, class, and age barriers, but they are also evident in access to resources like land.

The community comprises two organisations and related groups of people, namely Tyisa Nabanye[4] and the Melting Pot. Both are mainly working together. Tyisa Nabanye runs the community garden and is dedicated to small-scale intensive agriculture following the principles of permaculture. According to Catherine, one of the members, they are "seeking to improve food security, promote sustainable livelihoods, and create employment for our members" (Nicks, 2014). Members promote a sustainable and healthy lifestyle, which might be coined a counter movement to consumerism and make fresh organic produce available. The Melting Pot is an open-minded culture group and platform comprising Cape Townian artists and musicians. It hosts musical events, theatre plays, and art performances. Over the last two years, some members became active in "edutainment" – combining education and entertainment – and thus making some income for instance through music classes (The Melting Pot Africa, 2021).

One of the key activities is the weekly informal market – supported by Tyisa Nabanye and the Melting Pot – where community members sell fresh produce and host performances by different artists. During the last years, the market attracted many affluent visitors and seems to become one of Cape Town's new tourist attractions (cf. Lonely Planet, 2021; Nicks, 2014). In this vein, the community has been covered widely in the media.

When considering the ongoing land redistribution debate in South Africa, the initiative certainly sheds light on the issue of exclusive resource access and living conditions in metropolitan areas. The location on prime land in combination with the rapid growth of the city and its increasing attractiveness for tourists fuelled the debate with the state and developers tremendously over the last years. The government made several attempts to close the farm and sell it. It was only in 2016 that the last attempt failed thanks to local support and issues over land ownership between the municipality and the South African Department of Defence (Gee, 2017, p. 22). Erf 81 is still in the possession of the South African National Defence Force, and there are ongoing discussions about their willingness to donate the site (cf. Dart, 2019, p. 20). In general, one can observe escalating real estate prices and rents in this location over the past years. In the case of an eviction, it would not be possible for the community to maintain their activities in another spot located so close to the centre. There is no alternative plot available, and several community members would not be able to pay rent for an apartment in this area. Against this backdrop, the group has been confronted with a conflict. On the one hand, they face a certain pressure to act in favour of the government, hoping to get official access for free. On the other hand, they feel left alone and ignored in their land demands for community purposes. This will be illuminated further in the following sections.

South Africa's unequal agri-food system and Wright's lens of eroding capitalism

Overall, the Erf 81 community can be understood as an activist response to and kind of an emergency strategy in a highly unequal society and food system. I will elaborate on that in the following and use this as a starting point for the analytical lens of "eroding capitalism" which allows for a critical reflection and discussion of transformative attempts introduced by the Erf 81 community.

South Africa's past of racial segregation and related inequalities is still visible in many ways. For instance, black South Africans are more likely to struggle with poverty, low food insecurity, access to land, and the employment market (cf. Cousins et al., 2018; Statistics South Africa, 2018b). It is impossible to understand the challenges in South Africa today without looking back. Societal division, particularly along lines of race and class, can be traced back to apartheid, periods of colonialisation, and modernisation. Dynamics of de-agrarianisation, mainly industrialisation and promotion of large scale agriculture, led to a decline of the agricultural labour force and small scale farming and fuelled urbanisation (Du Toit, 2018). In part, this contributed to high un(der)employment and vital informal sectors, which contributed to precarious working conditions, insufficient salaries, and a crisis in social reproduction (Cousins et al., 2018; Du Toit, 2018). In Shivji's sense, labour exploitation subsidises capital and restricts necessary consumption (social reproduction) and also fuels self-exploitation (Shivji, 2017, p. 1011; Zhan & Scully, 2018). In the vein of neoliberalisation, westernised diets and unhealthy food have become more available and accessible for marginalised population groups. These developments are part of the country's integration into the globalised and industrialised agri-food system. Trade liberalisation in food, a shrinking natural resource base, displacement of small-scale farmers, overproduction, hunger, and volatile food prices are some of the features and dire consequences of the prevailing agri-food system, which are also visible in South Africa (Holt-Giménez, 2019; McMichael, 2013). However, in response to that, many food and agriculture movements engage with the "undemocratic architecture of the state system, its erosion of social and ecological stability, and its politically, economically and nutritionally impoverishing consequences" and thus refuse in one way or another the essentialism of neoliberalism, the market economy of capitalism, and the expression of a particular kind of class structure (McMichael, 2016, p. 654; Wright, 2019, p. 4). In this context, a CFI like the Erf 81 community seems to provide diverse solutions in a concrete jungle where access to healthy food, productive land, and housing poses challenges to those with tight budgets. Eric Olin Wright's advanced work on anticapitalism offers an interesting lens to contrast such CFIs' work and illuminate strategies to change the prevailing social system.

Whilst Wright admits that "[it] may simply be impossible to have a coherent strategy for the emancipatory transformation of something as complex as a social system," he is inspired by the destructive characteristics of capitalism, for example, minimal regulation of the market, exploitive working conditions, and unequal access to resources, to carve out five strategies that, in combination, create the potential for its erosion (2019, p. 37). In this way, they feed into the critical reparative approach. The five strategies for anti-capitalist struggles include: smashing capitalism, dismantling capitalism, taming capitalism, resisting capitalism, and escaping capitalism, which are illustrated in Table 4.1. He differentiates them into two categories: neutralising harms (column 1) and transcending structures (column 2).

Table 4.1 Typology of anti-capitalist strategies

	Neutralising harms		*Transcending structures*
The state		Eroding capitalism	**Smashing**
	Taming		**Dismantling**
The civil society	**Resisting**		**Escaping**

Source: Adapted from Wright, 2017, p. 16, 2019, p. 56.

The first one implies counteracting harm in the system. Whereas the second category comprises strategies that change and go beyond existing structures. Wright moreover defines the main actor introducing these strategies, for example, the state or civil society. Considering CFIs, it is thus important to particularly zoom in on the latter and the strategies of resisting and escaping capitalism. I briefly explain these strategies in the following section.

There are three strategies which require strong action of the state: taming, smashing, and dismantling. The most challenging and at the same time most revolutionary strategy is "smashing capitalism." It refers to a rupture in the existing system and thus destroying the dominant powers. Considering the difficulties in achieving this, Wright points out the rare successes in the past (2019, p. 40). Another related strategy is "gradually dismantling capitalism and the building up of the alternative through the sustained action of the state" a kind of a state-directed reform (Wright, 2019, p. 43). Both smashing and dismantling capitalism imply replacing capitalism gradually by socialist structures. Hence, these strategies are presented under the label "transcending structures" in Table 4.1. While this might be very difficult for social movements and partly requires interaction with/confrontation of the state, several scholars encourage food grower movements to change the city in a truly revolutionary and radical way which may imply an overhaul of society, a revolution to reclaim the city, making the state obsolete and building autogestion (e.g. Purcell, 2014, p. 150; Sbicca, 2014). Still, it seems impossible for mostly localised, small groups of food producers to initiate such a rupture. The third strategy is "taming capitalism"; it again implies powerful interventions of the state In contrast to the previously introduced ones it refers to building institutions which are capable of neutralising capitalisms harms. This may include regulatory mechanisms and redistribution. Certainly, it would also be difficult for representatives of civil society to engage with this strategy.

The remaining two strategies, however, directly refer to possible revolutionary attempts initiated by civil society. According to Wright, the strategy of "resisting capitalism":

> may not be able to transform capitalism, but we can defend ourselves from its harms by causing trouble, protesting. . . . This is the strategy of many grassroots activists of various sorts: environmentalists who protest toxic dumps and environmentally destructive development; consumer movements that organize boycotts of predatory corporations.
>
> (2019, p. 50)

The strategy is mostly applied by social movements and through resistance in everyday life, which could also be introduced by individuals. Resisting capitalism implies neutralising its harms and thus withholding efforts to benefit the capitalist class. For example, food movements are often critical of the dominance of corporate power in the retail sector related to low-quality food and inadequate prices. In response, these groups grow their own food. This less radical strategy is a within-system form of contestation and thus rather reformative; it thus parallels everyday forms of resistance (O'Brien, 2013; Scott, 1976). This understanding is in line with Bayat's and Biekart's rather optimistic thoughts; they emphasise that, "the neoliberal city is not a one-way street where capital reigns exclusively. Constraints notwithstanding, there are also opportunities for new actors and constituencies to make their mark on the configuration of urban life" (2009, p. 823). Hence, the interventions of CFIs might open windows of opportunity to criticise the dominant system and to build alternatives, for instance through solidarity networks.

Directly related to that, social movements and critical citizens may engage in what Wright calls the "escaping capitalism" strategy. This could comprise attempts for self-sufficiency and subsistence with an intention to transcend existing structures in rather isolated ways (Wright, 2019, p. 52). Marcuse for instance points to "sectors of everyday life that are free of capitalist forms, operating within the capitalist system but not of it . . . those are the sectors of the economy and of daily life that are not operated on the profit system, that . . . rely on solidarity, humanity" (2009, p. 195). It is the task of the following sections to use these analytical categories to illuminate the transformative attempts of the Erf 81 community in a predominantly capitalist system. In addition, Wright's strategies, in combination with the critical reparative approach, let us uncover inherent troubles and possible ways to overcome them. It is thus key to keeping in mind that CFIs' spheres of action are complex, and attempts to create broader changes and alternatives are often limited and intertwined with diverse challenges, as suggested by the *critical reparative approach*.

Uncovering inherent troubles: from self-responsibility to neoliberalisation?

In general, the Erf 81 community applies the strategies of "resisting capitalism" and "escaping capitalism." The following section of the paper provides case study detail to these strategies. Clearly, the lived realities of the members can be considered as a critical response to the dominant agri-food system and the larger unequal social system. I show that many of these revolutionary attempts sound promising but reveal considerable challenges and possible dangers. A critical reflection of the Erf 81 community and their progressive work shows that broader attempts to resist and change the prevailing agri-food system are likely to pave the way for neoliberal urban developments and division in the community. I frame these under the term of neoliberalisation which mirrors the component of trouble. One of the tasks of the *critical reparative approach* is to engage with these challenges and thus to suggest possible ways they might be overcome; the latter is part of the subsequent part, "Exploring potential hopes: Fights on diverse fronts beyond isolated niches."

The community's everyday practices mirror the strategy of *resisting capitalism*; particularly through criticism of large-scale agriculture related environmental degradation, increasing supermarketisation and commercialisation, decline of informal neighbourhood shops and food stalls, and the increasing availability of cheap and unhealthy food. They reveal this critical stance through self-responsibility, particularly the promotion of local and agroecological production including permaculture, community networks and skills, direct market structures, healthy diets, and sharing practices, for example, with seeds, knowledge, and produce (cf. Nicks, 2014; Mgcoyi, 2018, 2020). Many young visitors are attracted by these activities and related inherent resistance; for instance, they attend workshops or visit the community market. Beyond that, get inspired to grow their own food and increasingly value nature. The community is well aware of their popularity, and one of the young food growers proudly states, "We are making agriculture sexy again" (Petrie, 2016). Furthermore, these CFI's mundane forms of protest comprise land occupations. The issue of uncertain land access, related pending land requests of the community, causes them to maintain a precarious, illegal status which comprises critique of the government's profit-oriented land policies and unfulfilled promises of land redistribution and eventual donation. According to Wright, such endeavours and everyday protests may neutralise the harms of capitalism (2019).

Furthermore, the community's efforts in part also mirror the strategy of *escaping capitalism*. The Erf 81 community strives for their own utopia, a self-subsistence island, which again comprises the intention of self-responsibility. Some also refer to it as an eco-village, in the middle of Cape Town (Mgcoyi, 2018). It is a dream to be independent from supermarkets and exploitive employment structures. On a whole, the community's work seems to contribute to a kind of social and solidarity economy; social and democratic values are considered important, which apparently is different from an individualised interest in profit-making. From a theoretical perspective, these rather radical actions may contribute to ideally transcending capitalist structures. Still, one has to admit that this kind of isolation from larger capitalist structures is very difficult. Although the outlined strategies may sound hopeful, a closer investigation reveals troubles.

Throughout their attempts to create something revolutionary within the prevailing exclusive system, the community has acquired and acts in a certain position of privilege in combination with the potential for trouble. I apply the term privilege since not every CFI has the opportunity to use such a big piece of land in the city centre for free, run a weekly public market, and receive wide public attention. It might be termed volatile privilege considering the persistent challenges of eviction and the decay of the farm. Still, their work opens up a unique transformative space in a city struggling tremendously with historical, socio-economic, and political divides. This kind of privilege calls for responsibility in line with their aims for instance of solidarity, inclusion, and community-orientation beyond individual profit (cf. Bowness & Wittman, 2020). Clearly, the task at hand is not to deepen any of the existing divides. These aspects are certainly considerable challenges, as will be outlined in the following.

Like many other CFIs Erf 81 community struggles with an uneven social fabric, which can be traced back to its formation and its diverse personalities (e.g. Gee, 2017). The farm was originally set up by André, and still his presence seems to place him as the natural leader. However, the group consists of a variety of different, mostly younger individuals, many seemingly with their own agendas. It can easily be felt as an outsider, that slight conflicts over different intentions and strategies are existing, which in part is related to the tremendous diversity of the group. As highlighted earlier, many inhabitants have come from difficult backgrounds, for example, they grew up in deprived settings. Several members have no other place to stay and live on the farm, whereas others stay in the neighbourhood, and a few also live in townships at the fringes of the city. Some invest most of their energy into the weekly market, gardening workshops, as well as activities attracting many external guests, income, and visibility in the media. While they resist broader capitalist structures, they somehow seem to be tempted by profit and a bit of fame. Whereas a few individuals have a stronger focus on internal activities including ecological food production, and protection of partial subsistence. These people can be considered as more idealistic in terms of ecological sustainability and goals of food sovereignty and less celebratory in terms of exposing their work to a larger audience and creating income. Certainly, these experiences contribute to a fragmented community.

The outlined troubles in the context of the diverse backgrounds and interests of the members are closely intertwined with the broader reality of a complex capitalist system, which bites and shapes the actions of these revolutionaries. Marit Rosol emphasises that urban agriculture can be considered as an expression of the roll-back of the state and implies a "neoliberal ethos of self-responsibilization" (2012, p. 251). This parallels with Jamie Peck's ideas of the "shadow state" and ways governments are not taking responsibility to care for citizens, instead leaving communities to care for themselves (cf. Peck, 2003). This resonates with the experiences of several group members. They have not been fully integrated into the labour market; their salary does not suffice to make a living. Thus, they try to make a bit of money with the activities on the farm (e.g. gardening classes, children's groups, market, cultural performances). In addition, they rely on food that is produced on the farm, and food production has become an indispensable means of social reproduction (cf. Shivji, 2017). This context makes it difficult to maintain pure idealism for solidarity, sharing culture, and non-profit orientation over time. In addition, some more affluent members are in favour of more exposure and lifestyle marketing. It was in late 2016, that I visited the market with a small group of researchers from the University of the Western Cape. Our group of external observers, most of them Cape Townians, agreed that there is a high chance and simultaneous risk of becoming Cape Town's new tourist hot spot and "hipster place." This observation was supported by respective media coverage promoting the farm and the market as one of the city's tourist highlights, and self-marketing of the farm on Facebook and Instagram (cf. Cape Town Magazine, 2015). Against this background, the community strongly attracts privileged outsiders for instance tourists. They can easily access the hilly terrain by car, can afford the products on offer, which are mostly pricy vegetables, drinks, food, and second-hand clothes, and are familiar with the indicated social media channels. Those people do not necessarily share their ideals and rather contribute to a profit-oriented place.

More recent developments and activities in the community reveal a continued conflict between non-profit and solidarity orientations and more exposed start-up businesses intentions that differ from their original aims. In 2019, one of my former students from Germany visited the farm during her holidays and was inspired by the peoples' openness, whilst she enjoyed yoga classes and sitting in front of the fire. The Erf 81 market is still running and is now also listed in the travel guide Lonely Planet under Cape Town category "shopping" (Lonely Planet, 2021). In addition, the Melting Point offers the aforementioned edutainment. Hence, the audience attracted to the activities of the community remains exclusive. Originally, this space was not meant to be commercialised. However, up-scale coffee machines, fancy fruit blenders, pricy souvenirs, food stuff, and second-hand clothes tell different stories.

It was mainly in 2017, when flourishing market activities, free residence on state-owned land, and high public visibility further fuelled clashes with the government, which thought to close the farm down. In general, the government has been interested to sell the land to real estate investors for a long time. Throughout the years, further investors and city developers got attracted by the growing image of a hipster place in increasingly appealing neighbourhoods. It is hard to tell what most of the community members are striving for. However, it is clear that ongoing developments contribute to gentrification including upscale food provision. At the same time, the overall condition of the farm is increasingly decaying, as reflected in abandoned houses and vegetable patches. Consequently, the community had become cautious of outsiders, especially those wanting to report on their work, and activities on the farm felt less vibrant (e.g. Dart, 2019). One may argue that the community's position of privilege contributed to less beneficial aspects, for instance, too much public attention, a plethora of uncoordinated activities, and profit orientation of some members. The intensive debate with the municipality and the outlined troubles have provided a purpose for the community to organise themselves better and to set goals collectively (Gee, 2017). However, several years later, several activities and organisation on the farm still feels uncoordinated, spontaneous, and informal, and in part are driven by individual interests.

In sum, it is extremely important that the community dismantles the status quo and makes rather destructive and exclusive changes within their revolutionary intentions to avoid closure and falling-apart. I consider the advice provided by Andrew Zitcer in the context of a food cooperative in Philadelphia as helpful: "Overcoming the paradox of exclusivity requires efforts towards affordability, accessibility and reflective practice in order . . . to realize their transformative social and economic potential" (2015, p. 812). At this point, it is key to encourage internal critical reflection, reversion to original values, team building, and strategising.

Considering the community's volatile privilege, I suggest to "get their house in order" and will outline related potential avenues of hope in the next part.

Exploring potential hope: fights on diverse fronts beyond an isolated niche

This part shows that a combination of different strategies beyond the outlined strategies of resisting and escaping capitalism, may allow to achieve more idealistic and lasting changes. A burning question that is often raised in the context of CFIs is: Are these initiatives "niches" within the prevailing system or have they already managed to replace current structures? Kaika, for instance, is very sceptical and considers them as "immunologies" – ways to alleviate the symptoms of the current system without changing it (2017, p. 98). Certainly, self-help mechanisms and profit-oriented attempts – as outlined earlier – seem unable to fight inequality enshrined in an agri-food system dominated by capitalism. It is this experience that makes it important to suggest further avenues of hope in the context of the Erf 91 community. Against this backdrop, I propose two kinds of cooperation to strengthen the group's work and aims: (1) alliances with food sovereignty organisations and (2) alliances with scholar-activists. External, critical feedback and experience as well as strong coalitions could be helpful in re-balancing and strengthening truly revolutionary attempts.

First, strong (political) alliances could provide an opportunity to engage in further anticapitalistic strategies namely smashing/dismantling and taming capitalism. So far, the community has no clear and continued aims and standing politically. Hence, I recommend to (re)vitalise coalitions, particularly around common values (cf. Wright, 2019, p. 7), which also implies to keep the reality "beyond the fence" in mind and literally fighting on different fronts – radically and progressively at the same time (Siebert, 2020). I am referring to societal issues beyond the community, thinking, for instance, about racialised inequality in access to healthy food, education, and prime land in cities. Cooperation is essential to reap wider changes. So far, some community members clearly take sides politically – they are in favour of food sovereignty and are supporting South Africa's largest food sovereignty movement, the SAFSC (Mgcoyi, 2020; Petrie, 2016). Clearly, progressive and radical efforts towards food sovereignty chime in with strategies of both resisting and escaping capitalism. Throughout manifold activities of the community, first linkages with other non-profit organisations and progressive agri-food movements have been built; these are Slow Food Youth South Africa, the SAFSC, the Oranjezicht City Farm, and the Surplus People Project.

The work of food sovereignty organisations and campaigns in South Africa could play an integral role in initiating further civil society-led strategies in interaction with the state. For instance, the SAFSC engages in "resisting capitalism" from below, for example, through protests against bread corporations (South African Food Sovereignty Campaign, 2015). Beyond that, SAFSC and the German Friedrich Ebert Stiftung published the People's Food Sovereignty Act in 2018, which they seek to advance further, for incorporation into government policies.

In addition, the NGO Surplus People Project has fought several court cases for land access for smallholders in the Western and Northern Cape in the past. Efforts like these illustrate the use of legal frameworks and might shape state policies. In this context, it is key to keep in mind the radical politics required in the struggle for food sovereignty, including redistributive land reform and community rights to water and seed (Holt-Giménez & Shattuck, 2011, pp. 117–118). This chimes with the strategy of "taming capitalism" by using "state power to neutralize the harms of capitalism" (Wright, 2017, p. 13). Eroding the dominant agri-food system requires a diversity of continuous resistances and alternatives. If firmly articulated by civil society, advanced by different actors including the state and the market, and targeted from different angles, exclusionary food structures could be challenged (Borras et al., 2015, p. 613). Closer cooperation of the Erf 81 community with these organisations may help to specify their broad aims and tactics, and thus to hold the government into account, secure and establish a mandate and support for long term land access, inclusionary housing, community healing functions, for example, green spot in the city for recreation, urban gardening and healthy, affordable food provision, just to name a few aspects.

Second, for the Erf 81 community, scholar-activists could be key in realising and reflecting their troubles and strengthening their transformative efforts. The role of scholar-activists in wholeheartedly supporting the oppressed – reaching from deprived mine workers to urban squatter farmers – has gained more and more attention during the last decade (e.g. Edelmann, 2009; Piven, 2010). This kind of cooperation between researchers and those fighting at the forefront builds from shared values for broader social change and usually a sustained commitment for struggles. According to Wright (2019), shared basic anticapitalist values comprise three main pillars: equality, democracy, and community, which may build a first foundation.

In the case of the Erf 81 community, so-called engaged scholars could play a vital role in fighting at their side, creating critical consciousness, spreading the initiative's interests, and proposing future interventions and planning of the abandoned land. For instance, well-informed and reputable scholars could be helpful in proposing and promoting inclusive land use plans. Based on discussions with community members, research on the community has already attracted undergraduate and graduate students with a focus on urban food production and urban planning (e.g. Dart, 2019; Gee, 2017). While several students seemed to be keen in participating and learning about the initiative, there remains little long-term cooperation, precise feedback, or recommendations for the community so far.

Commitment as a scholar-activist starts with familiarity with the realities and struggles of the oppressed. This basically begins with "discovering and narrating what these projects are capable of producing" (Certomà & Tornaghi, 2015, p. 1127; Edelman, 2009). From my own experience, this mainly involves two issues: accountability and reciprocity. In line with Pulido's thoughts, a researcher attempts to reconcile his research interests with those of the community, which might also be referred to as "bottom-up accountability" (2008, p. 351; Hall et al., 2017, p. 4). Reciprocity "means looking for ways to reciprocate" (Pulido, 2008,

p. 352). Although my empirical research on Erf 81 is rather limited, I assume that further engagement with the group presents diverse opportunities. Especially participatory research and related direct actions with the community, for instance on resistance strategies, food sovereignty strategies, or social marketing could reap direct benefits for the community and interesting research results. I think, it is moreover one of the task of university lecturers, like myself, to create awareness about the reality of social movements including CFIs on the ground through teaching and to guide students when doing research in this field and encourage them to engage with these initiatives beyond pure data collection (cf. Montenegro de Wit et al., 2021).

Concluding remarks

This chapter asked how CFIs truly create hope through resisting the system in place and confronting related troubles. In this context, I highlighted the importance of carefully scrutinising CFIs transformative aims, potentials, and challenges. Applying a critical reparative approach helps identify inherent hopes and troubles and initiate respective support in the specific context. It is evident that deep roots and manifestations of poverty and inequality limit and challenge broader changes – as part of the notion of trouble. The City of Cape Town is considered as one of the most attractive metropolitan destinations for tourists; simultaneously, it is well known for its high inequality, drastic crime rates, increasing youth unemployment, and rapid growth. Against this backdrop, progressive and radical CFIs are doing essential work in making the city liveable and enjoyable for many and not only a few. At the same time, this setting confronts them with an additional responsibility to engage with historical, social, political, and economic divides. In this context, acting with sensitivity and responsibility is key. Certainly, it is one of the tasks of a CFI not to deepen inequality, for instance, through exclusive food markets. Wright's notion of "eroding capitalism" (2017, 2019) helps us to shed light on different transformative endeavours introduced by urban food growers, related restricting factors, and to identify potential further supporters.

Diverse layers of political action, such as everyday forms of resistance or overt protests, could be essential elements for change and fit the strategy of resisting capitalism and even escaping capitalism in Wright's typology. Starting with the lived experiences of urban producers helps avoid simplistic impact assessments of urban agriculture initiatives. The Erf 81 community certainly creates *hope* through their continued interest in creating "a space that serves the community" (Petrie, 2016). Their common aims like solidarity, community, sharing, food sovereignty, and ecological sustainability remain rather vague, but they attract many people, partly unite individuals with different backgrounds, invite cooperation with other progressive agri-food organisations, and create a wider public interest in their work. Furthermore, the case of the Erf 81 community illustrates how farming and related practices contribute to partial independence of the agri-food system in place. Daily actions dismantle inequalities and create alternatives. So far, the initiative has resisted eviction by the government. Certainly, some of the group's interventions are more radical than others, for example, land occupation and the informal market. Overall, the movement has been able to initiate change and critical debates on diverse fronts. However, more must be done to strengthen their group, define their aims, and carve out clear values to reap wider and lasting transformations in an unequal agri-food and social system. While applying Wright's typology is helpful to reflect local initiatives' resistance in the broader context of the social system, his work urges further intertwined actions in "resisting capitalism" with precise values. In addition, Wright calls for interventions in "taming capitalism" from above; cooperation with other progressive organisations is required in this specific case to interact with the state.

A recommendation derived from this experience is that targeting systemic inequalities in heterogeneous ways and on different fronts is key. This implies the application of different anticapitalist strategies, for instance contesting power, abolishing powers, and so on, through direct actions, mundane protests like doing food differently, uprisings, etc. (cf. Certomà & Tornaghi, 2015, p. 1124). The work of the Erf 81 community reminds us that it is key to act beyond a specific niche and self-responsibility, avoid isolation, and use the potential of networks with organisations and individuals, for example, scholar-activists, who can help in strengthening their aims and tactics. Simultaneously, CFIs are required to critically reflect on their work and ensure that their actions are still in line with their original intentions. This may also help to get their own "house in order" – in politicising their work and in carving out key values, to name just a few.

Wright's typology of anti-capitalist strategies helps to further nuance the work of the Erf 81 community. He emphasises, "What we need is an understanding of the anti-capitalist strategies that avoids both the false optimism of wishful thinking and the disabling pessimism that emancipatory social transformation is beyond strategic reach" (2017, p. 9). The applied critical reparative approach ties in here. Wright calls for a combination of diverse strategies for gradual change and multi-layered erosion. This cannot be done by one CFI or social movement alone. In this sense, the diversity of breaks created by variegated agri-food movements, activists, and related alliances could be the initial elements of eroding the prevailing and exclusionary agri-food system and thus challenging capitalism.

Notes

1 Malnutrition ranges from extreme hunger to obesity.
2 Erf is Afrikaans for plot.
3 According to the Declaration of Nyéléni (Nyéléni, 2007), it is defined as the "right of peoples to healthy and culturally appropriate food produced through sustainable methods and their right to define their own food and agriculture systems."
4 The Zulu name translated by the members as "feeding others" particularly refers to knowledge and food here. The non-profit and community gardening initiative was founded in 2013.

References

Alonso-Fradejas, A., Borras, S. M. J., Holmes, T., Holt-Giménez, E., & Robbins, M. J. (2015). Food sovereignty: Convergence and contradictions, conditions and challenges. *Third World Quarterly*, *36*(3), 431–448.

Bayat, A., & Biekart, K. (2009). Cities of extremes. *Development and Change*, *40*(5), 815–825. https://doi.org/10.1111/j.1467-7660.2009.01584.x

Borras, S. M., Franco, J. C., & Suárez, S. M. (2015). Land and food sovereignty. *Third World Quarterly*, *36*(3), 600–617.

Bowness, E., & Wittman, H. (2020). Bringing the city to the country? Responsibility, privilege and urban agrarianism in Metro Vancouver. *The Journal of Peasant Studies*. https://doi.org/10.1080/03066150.2020.1803842

Cape Town Magazine (2015). ERF 81 food market in Cape Town. *Cape Town Magazine*. www.capetownmagazine.com/events/erf-81-food-market-in-cape-town/11_37_55988

Certomà, C., & Tornaghi, C. (2015). Political gardening: Transforming cities and political agency. *Local Environment*, *20*(10), 1123–1131. https://doi.org/10.1080/13549839.2015.1053724

Cousins, B., Dubb, A., Hornby, D., & Mtero, F. (2018). Social reproduction of "classes of labour" in the rural areas of South Africa: Contradictions and contestations. *The Journal of Peasant Studies*, *45*(5–6), 1060–1085.

Edelman, M. (2009). Synergies and tensions between rural social movements and professional researchers. *The Journal of Peasant Studies*, *36*(1), 245–265.

Dart, T. (2019). *Reclaiming landscapes: The alternate vault of Tamboerskloof magazine, Cape Town, South Africa*. University of Johannesburg.

Du Toit, A. (2018). Without the blanket of the land: Agrarian change and biopolitics in post–apartheid South Africa. *The Journal of Peasant Studies*, *45*(5–6), 1086–1107.

Gee, W. (2017). *Motivations and outcomes of urban agriculture between different income groups in Cape Town, South Africa* [Master thesis, Universiteit Utrecht].

Hall, R., Brent, Z., Franco, J., Isaacs, M., & Shegro, T. (2017). *A toolkit for participatory action research*. Retrieved October 23, 2020, from www.tni.org/files/publication-downloads/a_toolkit_for_participatory_action_research.pdf

Holt-Giménez, E. (2019). Capitalism, food, and social movements: The political economy of food system transformation. *Journal of Agriculture, Food Systems, and Community Development*, 1–13. https://doi.org/10.5304/jafscd.2019.091.043

Holt-Giménez, E., & Shattuck, A. (2011). Food crises, food regimes and food movements: Rumblings of reform or tides of transformation? *The Journal of Peasant Studies*, *38*(1), 109–144. https://doi.org/10.1080/03066150.2010.538578

Kaika, M. (2017). "Don't call me resilient again!": The new urban agenda as immunology . . . or . . . what happens when communities refuse to be vaccinated with "smart cities" and indicators. *Environment and Urbanization*, *29*(1), 89–102. https://doi.org/10.1177/0956247816684763

Lonely Planet (2021). *Erf 81 market*. www.lonelyplanet.com/south-africa/cape-town/shopping/erf-81-market/a/poi-sho/1508515/355612

Marcuse, P. (2009). From critical urban theory to the right to the city. *City: Analysis of Urban Trends, Culture, Theory, Policy, Action*, *13*(2–3), 185–196.

McMichael, P. (2013). *Food regimes and agrarian questions*. Agrarian Change & Peasant Studies, 2. Fernwood Publishing.

McMichael, P. (2016). Commentary: Food regime for thought. *The Journal of Peasant Studies*, *43*(3), 648–670. https://doi.org/10.1080/03066150.2016.1143816

The Melting Pot Africa (2021). *About us*. Retrieved February 28, 2021, from http://themeltingpotafrica.com/

Mgcoyi, C. (2018). *Holding our ground: Voices for food sovereignty*. Retrieved March 4, 2021, from https://13africanfarmers.com/chuma

Mgcoyi, C. (2020, July 28). Find, make, use: Food, shelter and safety in the city. *Food Dialogues*. Retrieved March 4, 2021, from https://capetown.fooddialogues.info/talks/find-make-use-food-shelter-and-safety-in-the-city/

Montenegro de Wit, M., Shattuck, A., Iles, A., Graddy-Lovelace, G., Roman-Alcalá, A., & Chappell, M. J. (2021). Operating principles for collective scholar-activism: Early insights from the agroecology research-action collective. *Journal of Agriculture, Food Systems, and Community Development*. https://doi.org/10.5304/jafscd.2021.102.022

Nicks, C. (2014). Tyisa Nabanye – urban farm in Cape Town, South Africa. *City Farmer News*. Retrieved June 21, 2020, from https://cityfarmer.info

Nyéléni (2007). Synthesis report Nyéléni 2007 forum for food sovereignty. *Forum for Food Sovereignty*. Retrieved January 6, 2016, from www.nyeleni.org/IMG/pdf/31Mar2007NyeleniSynthesisReport-en.pdf

O'Brien, K. J. (2013). Rightful resistance revisited. *Journal of Peasant Studies*, *40*(6), 1051–1062. https://doi.org/10.1080/03066150.2013.821466

Peck, J. (2003). Geography and public policy: Mapping the penal state. *Progress in Human Geography*, *27*(2), 222–232. https://doi.org/10.1191/0309132503ph424pr

Petrie, O. (2016). *A farm in the rainbow – ERF 81 and Tyisa Nabanye*. YouTube. Retrieved March 4, 2021, from www.youtube.com/watch?v=ZRCSIe5Qwu4

Piven, F. F. (2010). Reflections on scholarship and activism. *Antipode*, *42*(4), 806–810. https://doi.org/10.1111/j.1467-8330.2010.00776.x

Pulido, L. (2008). FAQs. Frequently (un)asked questions about being a scholar activist. In C. R. Hale (Ed.), *Engaging contradictions: Theory, politics, and methods of activist scholarship* (pp. 341–365). University of California Press.

Purcell, M. (2014). "Possible worlds": Henri Lefebvre and the right to the city. *Journal of Urban Affairs*, *36*(1), 141–154. https://doi.org/10.1111/juaf.12034

Rosol, M. (2012). "Community volunteering as neoliberal strategy?" Green space production in Berlin. *Antipode*, *44*(1), 239–257.

Sbicca, J. (2014). "The need to feed": Urban metabolic struggles of actually existing radical projects. *Critical Sociology*, *40*(6), 817–834.

Scott, J. C. (1976). *The moral economy of the peasant: Rebellion and subsistence in Southeast Asia*. Yale University Press.

Shivji, I. G. (2017). The concept of "working people." *Agrarian South: Journal of Political Economy*, *6*(1), 1–13.

Siebert, A. (2020). Transforming urban food systems in South Africa: Unfolding food sovereignty in the city. *The Journal of Peasant Studies*, *47*(20), 401–419.

South African Food Sovereignty Campaign (2015). *About campaign*. Retrieved September 11, 2020, from www.safsc.org.za/our-story/

Statistics South Africa (2018a). *Quarterly labour force survey*. Pretoria. Retrieved June 3, 2020, from www.statssa.gov.za/publications/P0211/P02114thQuarter2018.pdf

Statistics South Africa (2018b). *Men, women and children findings of the living conditions survey 2014/15*. Retrieved June 3, 2020, from www.statssa.gov.za/publications/Report-03-10-02%20/Report-03-10-02%202015.pdf

Tornaghi, C. (2017). Urban agriculture in the food-disabling city: (Re)defining urban food justice, reimagining a politics of empowerment. *Antipode*, *49*(3), 781–801.

United Nations Population Division (2018). *World urbanization prospects: 2018*. Retrieved January 23, 2021, from https://data.worldbank.org/indicator/SP.URB.TOTL.IN.ZS?locations=ZA

Wright, E. O. (2017). How to be an anti-capitalist for the 21st century. *Theomai Journal Critical Studies About Society and Development*, 35(1), 8–21.

Wright, E. O. (2019). *How to be an anticapitalist in the twenty-first century* (1st ed.). Verso.

Zhan, S., & Scully, B. (2018). From South Africa to China: Land, migrant labor and the semi-proletarian thesis revisited. *The Journal of Peasant Studies*, 45(5–6), 1018–1038. https://doi.org/10.1080/03066150.2018.1474458

Zitcer, A. (2015). Food co-ops and the paradox of exclusivity. *Antipode*, 47(3), 812–828. https://doi.org/10.1111/anti.12129

5 Good food for all? Navigating tensions between environmental and social justice concerns in urban community food initiatives

Marit Rosol

Introduction

With most people now living in cities, and cities being hotspots for intensive resource use, urban actors are acknowledging more and more the central role of cities in our food systems. Cities are emerging as sites of education about agri-food systems – and the need to change them –, of re-politicisation and protest, and of envisioning and enacting alternatives. Urban food demand is playing a key role in the food system, of both hope and trouble, and often, urban initiatives are currently driving the sustainability agenda in agri-food. Many urban residents increasingly wish to consume regional, seasonal, and healthier foods that are produced and traded fairly. They are troubled about the damaging ecological, social, and economic practices and impacts of the current dominant agri-food systems[1] (Mendes, 2017; Rosol & Strüver, 2018; Sonnino et al., 2019; for early accounts, see Pothukuchi & Kaufman, 1999; Koç et al., 1999).

Going beyond individual action, urban Community Food Initiatives – mostly civil society actors, but often in partnership with small businesses, farmers, and municipal authorities – try to address agri-food sustainability issues through actions such as food waste diversion initiatives, the creation or reviving of local food networks, and through education efforts. With their actions and campaigns, they are hoping to work towards an end to food waste, for re-connection with producers, and a renewed relationship with food in general. As such, food-related urban social movements (for urban social movements in general, see Mayer & Boudreau, 2012) are an inspiration for hope as they have much potential and many positive effects.

And yet, there are also crucial constraints that trouble such positive view: First, we find an emphasis on market-based instruments and consumer choice, which is often tied to an overemphasis on educating individuals as consumers (Friedmann, 2005; Goodman et al., 2012). Second, there is the "local trap", assuming that the local scale is inherently good (Born & Purcell, 2006; Allen, 2010). Finally, these movements often have a limited focus on selected aspects of current food systems challenges. Despite long-standing calls for integrated social and ecological transformations of our agri-food systems to make them just, equitable, ecologically sustainable, and economically viable, it seems, in practice, Community Food Initiatives frequently focus on either the environmental aspect – often in cooperation with

DOI: 10.4324/9781003195085-6

producers, supporting direct marketing initiatives, etc. – or the social justice aspect, often with a focus on food insecure (urban) consumers that cannot afford or otherwise lack access to good, that is, fresh, nutritious, delicious food.

This chapter will focus on this last constraint, based on an analysis of the literature and my empirical research in German and Canadian cities. The objective is not to provide in-depth cases studies but to show the prevalence of single-issue approaches and then present and discuss some possible ways to overcome them following a critical-reparative approach – which together can hopefully inspire further in-depth studies.

My empirical research in German cities since 2017 shows that urban food initiatives tend to focus on topics of sustainability and environmental protection (by supporting organic agriculture, limiting food waste, etc.). As a result, there is often a narrow focus on sustainability at the expense of questions of justice and equity. While there are good reasons for single-focus initiatives (not least the always limited resources and capacities of mostly volunteer-run organisations) and it is certainly not expected that one organisation can attend to all or even most food-related challenges, the limited discussion in the burgeoning German urban food initiative scene of food poverty and food-related unequal participation in social life is still concerning. These issues are much more extensively discussed in other countries, such as Canada (Dachner & Tarasuk, 2017; Levkoe, 2006), where at times, however, we may find a rather limited understanding of ecological sustainability requirements (Interview 35/2018, Montreal).

Advancing a critical-reparative approach (see introduction to this book) that has been the foundation for all of my work (see Rosol, 2010, 2018, 2020),[2] this chapter seeks to contribute to debates on tensions regarding the split between "environmental" and "social" framings or urban community food initiatives and how to overcome them. In what follows, I will first give an empirical introduction, briefly presenting examples from original empirical research in German and Canadian cities to illustrate the tensions mentioned earlier, thus troubling the idea of Community Food Initiatives. Second, I provide a short insight into the state of research on alternative food movements as well as on Household Food Insecurity, Community Food Security, and Food Justice. Turning to potential ways forward and hope, in the final section, I present two integrative and promising approaches to overcome these divisions and briefly discuss potentials and limitations: first, the Canadian concept of Community Food Centres (CFCs), and second, Food Policy Councils as a tool to envision and enact integrative food policies on a municipal level. While not shying away from critique and troubling these examples by pointing out difficulties and shortcomings, I will present these two examples as actually existing attempts to overcome divisions, and thus points of inspiration and hope.

Divergent motivations of urban food initiatives: Empirical evidence for the environmental versus social justice divide

Methods

This chapter is based on my empirical research in German and Canadian cities (Berlin and Frankfurt, Toronto and Montreal) in the summers of 2017, 2018 and

2019. Based on an extensive database on urban food initiatives for each of these cities I had created with the help of research assistants, I chose several food initiatives for a more in-depth case study (11 in Berlin, 5 in Frankfurt, 5 in Toronto, 7 in Montreal). The case studies were chosen for theoretical reasons (theoretical sampling), representing as much diversity in their aims, approaches, or target groups as possible, as well as the availability of previous research publications, giving preference to those that have been researched less extensively. My research and the findings presented in this chapter are based on data triangulation, specifically, the analysis of existing documentation on these initiatives (all kinds of published material, including academic literature as well as grey literature such as reports or newsletters), their websites and general media coverage, visiting the initiatives and/or public events where their representatives were present, as well as in-depth, semi-structured interviews with at least one representative of an initiative and with other food experts in the study cities (N = 39). Interviews were recorded, transcribed, and analysed with help of NVivo qualitative data analysis software. The presentations on these initiatives and other empirical findings refer predominantly to the time the empirical research took place (2017–2019) with some desk-based updates in 2021 when possible.

Examples from German cities

Overall, my empirical research in German cities since 2017 shows a tendency of the initiatives to focus on the topics of sustainability and environmental protection, while much less attention is given to food poverty, Household Food Insecurity, and unequal social participation in food-related activities. My research revealed three main areas of interventions: (a) initiating new or supporting existing Alternative Food Networks; (b) food waste prevention; and (c) creating Food Policy Councils. In this section, I give a brief introduction to some representative initiatives pursuing the first two strategies. Food Policy Councils will be presented and discussed in Section 4 as one possible way to overcome the environmental-social divide.

Alternative Food Networks (AFNs)

Many of the newly founded urban food initiatives in Germany support young or new, small-scale, and mostly organic farmers by establishing or promoting Alternative Food Networks (AFNs) with (urban) consumers. My case studies include those who seek to establish new marketing opportunities for small-scale farmers, for example, *Futterkreis*, a new Food Co-op in Frankfurt founded by a consumer group, or *Marktschwärmer*, a business model that combines a pop-up farmers market with an online ordering platform which was invented in France. Others seek to support farmers by enabling access to land for new and young farmers (e.g. *Ökonauten* and other land purchasing cooperatives as well as *Bündnis Junge Landwirtschaft* as a political advocacy group and mutual support network). Finally, some seek to support regional food systems by investing in various parts of the food chain, including production, processing, logistics, food preparation, and retail (*Regionalwert* AGs in several German regions, *Bürger* AG *für regionales und nachhaltiges Wirtschaften* in

Frankfurt) (for more information on AFNs in general as well as details on some of the initiatives mentioned, see Rosol, 2020; Kumnig & Rosol, 2021; Rosol & Barbosa Jr, 2021).[3]

Avoiding food waste

Another very strong arm of urban-based food movements in Germany focuses on food rescuing or food waste prevention, being troubled by current amounts of food being wasted. Taking a different approach to the already established *Tafeln* (comparable to food banks and in existence since the 1990s), I interviewed representatives from initiatives that opened restaurants which cook with what would otherwise be regarded as and turned into food waste (*Restlos Glücklich* in Berlin and *Shout Out Loud* in Frankfurt), established a new business that delivers "cosmetically imperfect" organic produce to canteens and other corporate clients at a discount (*Querfeld* in Berlin), and pick up and distribute discarded food items at supermarkets (*foodsharing.de*) (Morrow, 2019).[4] Some of them are organised as a civic association (*eingetragener Verein* – e.V., such as *Restlos Glücklich e.V.* in Berlin and *Shout Out Loud e.V.* in Frankfurt), others as a business (such as *Querfeld*). Most of them also entail an educational component – teaching the general public and/or specific groups (e.g. children in the case of *Restlos Glücklich*) about the issue of food waste and how to prevent it. Most of them position themselves generally within a sustainability framework. Many of these initiatives were inspired by the 2011 "Taste the Waste" documentary by a German filmmaker (who incidentally, later on, became one of the founders of the Cologne Food Policy Council).[5]

Examples from Canadian cities

In Canada, on the other hand, civil society initiatives and research projects, troubled by the uneven access to good food, have long been concerned with strategies to overcome food insecurity, that is inadequate access to food due to a lack of financial resources. While there are also many examples of AFNs in Canada – see, in addition to a wealth of literature, our own research on a newly founded producer co-operative that offers a harvest box style CSA in Calgary, Alberta (Rosol and Barbosa Jr, 2021) – more than in Germany, Canadian urban food initiatives are also concerned with food access, especially in underserved low-income neighbourhoods.

2.3.1 Food access/Food provisioning/Social restaurants

During my research in Toronto in 2017, I was able to visit and interview several projects that approach food in a context of social exclusion and poverty. I interviewed the founder of a social enterprise that is working towards bringing good food to underserved neighbourhoods in Toronto (*Building Roots*); one of the founders of *Community Food Centres Canada* (see section 4); the former director of the most important and oldest food organisation in Toronto, *FoodShare*, who is deeply concerned not only about good food but also about poverty and food insecurity; and a student food organisation working towards better access to good, healthy, sustainable food for students and on campuses (*Meal Exchange*).[6]

In 2018, I visited and interviewed inspiring projects in Montreal, like a social restaurant where everyone can eat a very affordable lunch in a dignified environment (*Resto Plateau*), a student-run collective kitchen that serves 500 vegan lunches a day by donation (*People's Potato*), a large food bank that, amongst other activities, distributes food donations to 288 community organisations, including those I interviewed (*Moisson Montreal*), and a local anti-poverty organisation that understands food as a human right and integrates local food systems efforts and food programming into their work (*Parole d'exluEs*).[7]

Environment versus social justice?

My research revealed a disconnect between initiatives focusing either on environmental goals – which often pay limited attention to questions of access, justice, and equity – or on food access and food provision, which do not always pay attention to how ecologically sustainable the food items they process or distribute are. The first version often appears in the form of Alternative Food Initiatives (AFI), the second has a focus either on food access/food programming or on advocating for income-based approaches to combat root causes of food insecurity. Generally speaking, sustainability-driven *AFIs* aim to transform the production, processing, distribution, and consumption of food. *Community food security initiatives* focus mostly on enabling access to (good) food for low-income or marginalised populations. Finally, practices and scholarship concerned with *income-focused strategies* to overcome food insecurity seek to shift our focus away from food towards income, public welfare, and cost of living. While Alternative Food Networks are sometimes criticised for their overreliance on market-based instruments and consumer choice, and thus for their social exclusivity, income- or food access-focused approaches to food insecurity may be seen as ignoring or not being troubled by the need to transform the current dominant industrial food system.

In what follows I give a brief introduction into the general literature of these distinct approaches that try to advance the issue from either an environmental (and often producer-focused) or a social (and often consumer-focused) perspective.

Environment – Alternative Food Initiatives (AFI)

Alternative Food Initiatives (AFI) generally aim at environmentally and socially responsible ways of producing, distributing, and consuming food (Allen et al., 2003; Guthman, 2008a; Levkoe, 2011, 2014; Rosol, 2020; Wekerle, 2004). However, in practice the emphasis is mostly placed on environmentally friendly food production and on securing the livelihoods of producers. Alternative food systems in the Global North are usually not initiated as a response to food insecurity (hunger), but to provide an alternative to the dominant industrial food system based in a critique of its environmental, social, and economic impacts (Alkon & Guthman, 2017; Guthman, 2008a; Levkoe, 2011). Alternative Food Networks (AFNs) specifically hope to provide spatial, economic, environmental, and social alternatives to conventional food supply chains (Goodman et al., 2013; Renting et al., 2003; Rosol, 2020; Whatmore et al., 2003). Typical distribution models include Farmers Markets and weekly box-schemes. They are usually based on Short Food Supply Chains (SFSCs), which in contrast to conventional supply chains involve fewer actors, shorter and more direct connections between producers and consumers, and reduced geographical distance between production and consumption. The absence of intermediaries means that participating producers are able to retain a larger part of the value created, enabling economic feasibility and securing livelihoods (Hinrichs, 2000).

AFNs have been criticised for their over-reliance on market mechanisms and for focusing too much on healthy and local, high-quality food for the educated and affluent consumer (Goodman et al., 2012), for their colour blindness and white universalism (Guthman, 2008b; Slocum, 2007), for neglecting the question of labour (Myers & Sbicca, 2015; Weiler et al., 2016), and for falling into the "local trap" (Born & Purcell, 2006; Allen, 2010). Most importantly, according to Guthman et al. (2006), it is still an unsolved question whether and how AFN instruments developed for alternative food provisioning and in support of small-scale farmers can possibly address food insecurity.

Social – the right to food, Household Food Insecurity (HFI), Community Food Security (CFS), Food justice

Food was first recognised as a fundamental *human right* in the 1948 Universal Declaration of Human Rights (Rideout et al., 2007). Availability, adequacy, and accessibility are defined as the three main elements of the Right to Food (UN Economic and Social Council, 1999). Experts also include the social and cultural function of food and its significance for an active participation in society (Ziegler, 2008, p. 9).

As one of the most important social determinants of health (Raphael, 2016), *Household Food Insecurity* (HFI) is relatively well studied in the Canadian context (for other OECD countries, see Riches & Silvasti, 2014). Household food insecurity is defined as "inadequate or insecure access to food due to financial constraints" (Dachner & Tarasuk, 2017, p. 222). The annual Canadian Community Health Survey, which since 2005 includes a Household Food Security Survey Module,[8] consistently shows a high and increasing number of food insecure households (Dachner & Tarasuk, 2017). Despite a well-demonstrated link between Household Food Insecurity and poverty (Dachner & Tarasuk, 2017), responses to food insecurity – in Canada and elsewhere – have generally focused on charitable food assistance, mostly in the form of food banks. Food banks, which are generally volunteer-run, and collect and distribute food donations, grew in Canada as temporary emergency measures to welfare cutbacks in the 1980s (ibid.). They have long been criticised for not addressing poverty as the structural cause of hunger; relying heavily on unpaid labour and unpredictable food donations; being inefficient, stigmatising, and demeaning; and shifting the focus from rights to charity (Poppendieck, 1998; McIntyre et al., 2015).

Nonetheless, in Germany and Canada the predominant approach to food insecurity is still charity- and emergency-based, handing out food items to the "needy", often combined with efforts for food waste diversion through food banks or *Tafeln*. Such an approach not only does eschew the root cause of food insecurity – lack of income – but may also be adding insult to injury. As a prominent food-as-human-right advocate, the director of FoodShare Toronto, Paul Taylor puts it: "Treating low income people like walking compost bins is not the solution to this issue" (CBC Radio, 2020).

Community Food Security (CFS) emerged in the 1990s from a critique of charitable food assistance. Defined as "a situation in which all community residents obtain a safe, culturally acceptable, nutritionally adequate diet through a sustainable food system that maximizes community self-reliance and social justice" (Hamm & Bellows, 2003, p. 37), it acknowledges structural causes of food insecurity. CFS initiatives use community food programming tools (community kitchens, "good food boxes"), but also advocacy and policy as important strategies to achieve CFS (Winne et al., 1997). However, according to Dachner and Tarasuk (2017, p. 229), like food banks, these initiatives cannot tackle the root causes of food insecurity as they cannot address income directly (see for an early critique also Tarasuk, 2001).

As mentioned earlier, Guthman et al. (2006) argue that in the United States the strategic alliance of community food security and sustainable agricultural movements (and the hope that by eliminating market intermediaries both small-scale farmers and low-income consumers would benefit) failed because farm income security and Household Food Security may be incompatible goals under capitalism. While there are differences between the countries regarding their agricultural and welfare policies, this incompatibility is likely equally true in Canada and Germany. The unsolved trouble between environmental and social justice concerns was one of the main reasons why the Community Food Security Coalition in the United States decided to disband in 2012 (Morgan & Santo, 2018, p. 36).

Food justice was developed in the United States mainly in response to the critiques of Alternative Food Initiatives as well as the failure of CFS initiatives just mentioned. Proponents demand that alternative food movements must meaningfully engage with issues of hunger, race, and class (see Swan, 2023, this volume), if they wish to create ecologically sustainable and socially just food systems. While recognising that food insecurity is related to poverty, Food Justice understands the causes of food insecurity *and* ecologically damaging food systems as rooted in broader systems of oppression, such as extractive capitalism, racism, patriarchy, and colonialism (Alkon & Agyeman, 2011). It thus combines the demand for access to good food for all, regardless of income, origin, and education with democratic participation and fair labour conditions throughout the entire food sector (i.e. in production, processing, retail, gastronomy) (Alkon & Agyeman, 2011; Allen, 2010; Gottlieb & Joshi, 2013). In practice, food justice has taken many forms, including programmes that provide access to local organic vegetables to low-income communities as well as urban agriculture programmes that empower marginalised communities.

According to Julie Guthman, however, most food justice organisations effectively limit themselves to access questions, and "a critique of general inequities across the food system is immanent at best" (Guthman, 2014). She goes on: "In practice, the food justice movement . . . gives scant attention to other injustices in the food system, particularly those arising in food production: exposure to toxic chemicals, poor working conditions as they apply to health and safety, and disparities in wages and employment." and concludes:

> The sources of injustice in the food system (e.g. privatized and racialized land access and tenure, immigration and border policy, inadequate health and safety regulation) are nearly invisible in the food justice discourse, which, in many respects, is simply building alternatives that comfortably exist alongside the industrial food system.
>
> (Guthman, 2014)

Writing from a Canadian context, Kepkiewicz et al. (2015) assert that food-based organisations led by marginalised and racialised communities or individuals have been more successful in enacting a comprehensive understanding of Food Justice, while organisations primarily lead by white, middle-class people have often failed to adequately understand and meaningfully enact their roles and responsibilities in relation to food justice goals. Importantly, they argue that Food Justice is not to be achieved by "including" people of colour in CFS or AFN projects. Rather, "food justice research and practice should . . . connect food system inequities to interlocking structures of oppression, such as capitalism, patriarchy, white supremacy, and colonialism . . . and think more carefully about the structures that create privilege and disadvantage in our society, and that damage our relationships with the land and each other" (Kepkiewicz et al., 2015, pp. 99–100). In the Canadian context, this means to specifically address the ongoing settler colonial project (which is not to be subsumed under institutional racism) and a stronger "consideration of food's intimate relationship to land" (Kepkiewicz et al., 2015, p. 103) and the "ongoing violence against Indigenous lands and food systems" (Kepkiewicz et al., 2015, p. 101).

These divergent assessments point to the need for further empirical research. It must be acknowledged of course that it would be difficult or next to impossible for one organisation to offer access to good food for marginalised populations *as well as* attending to all the injustices across the food system – not least because time, money, and personnel are usually especially limited and there are many constraints these initiatives are working under. Despite the difficulties of successful alliance-building mentioned earlier, this would call for focusing our attention on possible alliances between different organisations to jointly achieve the necessary social and environmental transformations.

Exploring integrative approaches

In this final section, I will present and discuss two hopeful ways of bridging different food-related concerns: first, the comprehensive approach by Community Food

Centres Canada, and, second, Food Policy Councils as integrative approaches and tools for urban food policymaking.

***Providing good food for all:* Community Food Centres Canada (CFCC)**

Partly in response to the inadequacy of food charity and the neglect of structural issues in community food programming, since 2012 Community Food Centres (CFCs) have offered hope through an innovative approach that addresses the problem of inadequate food access beyond the provision of emergency food. A Community Food Centre (CFC) is a place-based, physical space that "uses food as an entry point to promote the physical and emotional health of individuals and communities, and to develop community-based and state-level strategies to address challenges within the food system" (Levkoe & Wakefield, 2011, p. 250). The CFC model was developed after The Stop CFC in Toronto, which started out as a small urban food bank, but gradually evolved into a food and community hub (Saul & Curtis, 2013; Levkoe, 2003, 2011). The national NGO Community Food Centres Canada (CFCC) was founded in July 2012. Its mission is to provide resources to partner organisations across Canada "to create Community Food Centres that bring people together to grow, cook, share, and advocate for good food" (Mission Statement on website).[9] Good food is defined by CFCC as fresh, nutritious, and delicious, sourced in as socially and environmentally responsible a manner as possible.

CFCs seek to provide access to food in a dignified and respectful way (e.g. by eliminating long line-ups and intrusive means testing), while also being troubled about the ecological sustainability or nutritional quality. The model is based on the assumptions that food is a public good and not a commodity. CFCC also proclaims that change must happen at the governmental, community, and individual level. They hope to draw connections from food to broader social, political, and ecological issues. According to their dictum "Good Food Is Just the Beginning", they educate about the links between food insecurity, growing poverty, and insufficient welfare rates, and they push for policy-level, not charity, solutions (Levkoe & Wakefield, 2011; Saul & Curtis, 2013; Scharf et al., 2010). Overall, the CFC approach can be understood as a mainly food-based (not income-based) response to household food insecurity that acknowledges the inadequacy of and thus seeks to overcome charitable responses, exemplified by food banks.

The CFC work rests on three pillars: (1) healthy food access programmes; (2) healthy food skills programmes; and (3) education, advocacy, and engagement programmes. The third pillar is particularly relevant here as it has the potential to address structural food systems injustices. That at least one staff position in each CFC is dedicated to this third pillar should ensure that advocacy plays a role in their work and programming (Interview 12/2017 with co-founder of CFCC, Toronto). Besides specific staff positions and programming that support each of the pillars individually, the aim of CFC is also to bring the different objectives in conversation with each other and pursue objectives across their different programmes. One tool to achieve this is that hiring processes are to be guided by the main values and principles of CFCs,[10] codified also in partnership criteria and operating standards,[11] which includes a commitment to social justice values and an understanding of systemic and policy issues (Interview 12/2017, Toronto).

In practice, however, CFCs abilities to advocate for structural changes are limited, not least because of their charity status and their dependence on philanthropy. Also, in their day-to-day operation, immediate needs or specific circumstances of the local organisation (which run the local CFC in partnership with CFCC) often dominate. If and how this model indeed can bridge the social-environmental divide requires further empirical analysis. What the model certainly offers is a much more comprehensive approach to community food programming as well as a more dignified access to good, healthy food and this way also a possibly bridging environmental and social concerns.

Integrative urban food policies: Food Policy Councils (FPC)

Acknowledging the intrinsic limitations of individual actors and initiatives to solve societal problems as well as the difficulties in alliance building mentioned in the previous sections calls for exploring other avenues. A hopeful field for an exploration of how to bridge the social-environmental gap are efforts towards food policymaking at the local level because they – at least theoretically – exemplify integrative, cross-sectoral approaches (Moragues-Faus & Morgan, 2015; Morgan, 2015; Mendes, 2017; Sonnino et al., 2019; for early accounts see Pothukuchi & Kaufman, 1999; Koç et al., 1999). One interesting laboratory and tool for enacting food democracy and bridging environmental, democracy, and social justice concerns are Food Policy Councils (FPCs), to which I turn next.

Food Policy Councils in Canada

Canada and especially Toronto have a long and widely acclaimed history of FPCs (Blay-Palmer, 2009; Friedmann, 2007; Schiff, 2008). Important to our discussion here, the Toronto FPC, one of the first FPCs in North America, was founded in 1990 precisely with "a vision of food security based on both social justice and environmental sustainability" (Welsh & MacRae, 1998, p. 237). An important outcome of their work, the council-approved Toronto Food Charter (Toronto City Council, 2001a) and the related Action Plan (Toronto City Council, 2001b), not only call for a more sustainable food system but also comprehensively address food insecurity and social issues. The Toronto food charter calls for an integrated and multi-scalar approach. Besides enabling better access to healthy foods, it also lists a commitment to "advocate for income, employment, housing and transportation policies that support secure and dignified access to the food people need" (Toronto City Council, 2001a) and to work with other scales of government (i.e. the provincial and the federal level) to achieve these goals and address causes of hunger.

Food Policy Councils in Germany

In Germany, following North American models emerging since the 1990s, FPCs have only been established recently since 2016 (Doernberg et al., 2019; Hoffmann et al., 2019; Thurn et al., 2018). Most FPCs are strongly civil society driven, some with the formal participation of government officials and (small) business representatives. In April 2016, a few weeks after the first German Food Policy Council in Cologne, the *Ernährungsrat Berlin* was founded as the second German FPC. It is citizen-led and works towards "food democracy and re-localization" of food in the Berlin-Brandenburg region (Hoffmann et al., 2019). The *Ernährungsrat Frankfurt* (Main) was founded soon after as the third German FPC in August 2017. As of October 2021, there were 21 existing FPCs in Germany (most of them founded in 2018) and about 40 in the process of formation.[12] I conducted interviews with several members of the Berlin and the Frankfurt FPC.

By definition, FPC have an integrative approach – attending to many food-related policy issues and advising and/or expressing demands towards local governments. However, at the time of research, and thus the early days of German FPCs, the main focus of the FPCs in Frankfurt and Cologne, and to a lesser extent in Berlin,[13] was on enabling, promoting and supporting local and regional food networks and farmers/producers in their production and marketing efforts, reconnecting (urban) consumers with regional producers, environmentally friendly local food production, and healthy eating, often combined with food education/literacy work.[14]

I found less attention to and sometimes ignorance regarding social issues, access questions, or (food) poverty. For example, when asked, members of the FPC in Frankfurt, directed me to their education efforts (during a workshop in 2019), proclaimed that membership in a CSA would be actually quite affordable or that with a little extra effort one could buy, cook and eat organic food relatively cheaply (Interview 14/2017, Frankfurt),[15] or suggested that low-income people could produce their own organic food in a community garden (Interview 26/2017, Frankfurt)[16]– efforts and suggestions, which, while important and not problematic per se, are certainly not a substitute for attending structurally to questions of social justice and household food insecurity (HFI). Contrary to a common misconception I noted amongst German food actors, Canadian research has shown, moreover, that HFI is not a problem of insufficient food skills (shopping, planning, handling, cooking, gardening) (Huisken et al., 2017; Spoel & Derkatch, 2020).

Also the 2018 "Frankfurt Proclamation", adopted at the second congress of German speaking FPCs held in Frankfurt, which lists many of the distortions and damages of our current dominant food system and suggests solutions and voices demands for governments, mentions the word justice only in very general terms ("A democratic, just, free and cosmopolitan society is essential for the success of the food transition!") and does not explicitly name social questions, food access or food insecurity.

Based on my research, I interpret this lack of attention, first, as a result of the essentially self-selective recruiting of FPC members. Volunteering for a FPC naturally is more attractive to people who work on food issues and not necessarily to people who work more generally on income and poverty (which are the causes of HFI, see earlier). Importantly, it also rarely attracts those that have lived experience of poverty or marginalisation (for a similar observation see also Hoffmann et al., 2019).

The second reason is a general lack of awareness of the problem and prevalence of HFI by municipal governments, the broader German public, and also the German food movement. This may be in part caused by the lack of data on HFI. Possibly, the extent of HFI is also less severe compared to Canada due to a different welfare and retail system which generally makes food in Germany relatively affordable; precise data is missing. Furthermore, some food organisations in Canada offer services that can be taken for granted in Germany, such as an affordable school lunch as well as lunch options for university students on campus which in Germany is standard and usually provided by a non-profit organisation.[17]

Nonetheless, HFI is an issue in Germany and it is becoming more severe in the face of rising poverty levels (Pieper et al., 2020), affecting an ever-increasing part of the population (Augustin, 2020). Attending to the social question may actually be a crucial role for a FPC to play, complementing the expertise on sustainable food production and distribution that other actors bring to the table (also voiced in interview 43/2019, member of Berlin FPC, Berlin). As the FPC movement in Germany is still in its infancy, it remains to be seen how attentive and supportive FPCs will be to the needs of marginalised urban populations.[18]

Conclusion and outlook

In this chapter, I first briefly presented examples from original empirical research in German and Canadian cities to illustrate the tensions between environmental and social goals and, for Germany, the dominance of environmental concerns and a still rather limited attention to issues of food insecurity and food poverty. Second, I provided a short review of research on alternative food movements as well as household food insecurity, community food security, and food justice. Turning to possible ways forward and in the spirit of a critical-reparative approach, in the final section, I presented and discussed the potential but also the limitations, the "hope and troubles", not only of community food initiatives in general but also of integrative approaches in particular, specifically Community Food Centres Canada and Food Policy Councils, to overcome these divisions.

The need for an integrated approach to foster food systems that are democratic, ecologically sustainable, as well as economically and socially just, has been championed by scholars and grassroots movements for some time (Allen, 2010; Holt-Giménez & Shattuck, 2011; Power, 1998). It is integral, for example, to food sovereignty frameworks that identify underlying structural logics of capitalist appropriation, patriarchy, colonialism, and racism as drivers of both social and ecological inequities and harm (Alkon & Guthman, 2017; Gottlieb & Joshi, 2013; Wittman et al., 2010). Nonetheless, while there have been successful attempts of integration, one of the severest tensions within food-based social movements remains between "the cause of sustainability (where this is largely framed in environmental terms) and the cause of social justice (where this is framed in terms of advancing race, class and gender equity)" (Morgan & Santo, 2018, p. 36).

A second important ongoing debate is around whether the social question and especially HFI is better addressed through food related efforts and programming or through income-based approaches. For a long time now, many studies have cautioned that the Human Right to Food can only be realised if the roots of poverty are addressed and the erosion of welfare is reversed (Riches, 2018; Rideout et al., 2007). This perspective advocates for income-based responses, including poverty reduction strategies such as a guaranteed basic income, and changes to labour and wage conditions and regulations (Power et al., 2019; Tarasuk, 2017).

However, while household food insecurity is caused by financial constraints, there are other major troubles with the food systems that prevent access to good food for all. Although not developed as a response to hunger and certainly not being able to eliminate food insecurity, Alternative Food Initiatives (AFI) can play a role in community food security efforts (Wakefield et al., 2012). More importantly, AFI remind us that although food insecurity is caused by insufficient income and the high cost of living, it is also necessary to trouble the current food system. Beyond adequate income (a necessary but insufficient condition), access to good food for all also requires a sustainable, safe, and high quality food supply, based in just conditions for food producers across the globe (Welsh & MacRae, 1998).

Even if urban Community Food Initiatives have to focus on one specific aspect of food systems in their day-to-day work, they need to be attentive to both environmental and social questions. A focus on either environmental or social aims may retain analytical or strategic usefulness (e.g. to ensure respective visibility), and there might also be certain logical and political incompatibilities and trade-offs that may partly explain why these frictions remain so intractable. However, food also provides a unique and vital lens to explore the potentials for more integrative ecologically *and* socially just transitions.

(Urban) Community Food Initiatives can raise awareness, demonstrate alternatives, and cater to immediate basic needs, but they cannot solve the food system and food insecurity crisis on their own. To achieve food security in a comprehensive sense, to enable access to good food for all, we need to make the food system more sustainable and fairer – through food and agricultural change and policies – while we equally need to combat poverty and raise incomes – through economic and social change and policies.

Abbreviations

AFI = Alternative Food Initiatives
AFN = Alternative Food Networks
CFC = Community Food Centres
CFCC = Community Food Centres Canada
CFS = Community Food Security
HFI = Household Food Insecurity
FPC = Food Policy Council
SFSC = Short Food Supply Chain

Acknowledgements

I am indebted to the interviewees who shared their ideas and experiences with me. I thank the editors and an anonymous reviewer for their comments, and Natalie Bakko for her careful proofreading of an earlier version of this chapter. This research was supported through the Canada Research Chair programme of the Social Sciences and Humanities Research Council of Canada (SSHRC).

Notes

1 Our current agri-food systems cause severe environmental, economic, health, and social problems (for an overview, see Wiskerke, 2009; De Schutter, 2014; Galt, 2013; Friedmann, 1993; Holt-Giménez, 2017; McMichael, 2009). They are not only a leading cause of severe environmental degradation and climate change, but they are also already tremendously impacted by it (IPCC, 2019; Rockström et al., 2020). As the 2008 UN world agricultural report (IAASTD, 2009; Beck et al., 2016) concluded: "Business as usual is not an option".

2 In Rosol (2018), for instance, taking the example of community gardens, I argue that "a universalizing question, if urban community gardens are an expression of urban neoliberalism or of commoning, poses a misguided binary. In any case, this question can never be answered in the abstract but requires thorough analysis of these projects and their context on several scales. The interesting aspect about community gardens is precisely that they are not either/or, but always have both potentials." Instead, I suggest focusing on "the how and where of specific projects to better understand its functions, forms, tensions and contradictions – that more often operate on different scales and argue for detailed empirical analysis of both specific projects and the general context of their activities. This means, for example, to ask about their aims, practices, their reflections on their own acting and the context they work in, as well as the effects of their doing." I argue for careful critique because unlike others who caution us against the potentially counterproductive effects of critiques of community food initiatives' entanglement with neoliberalism, I believe these critiques are crucial for transformative change. I conclude with emphasising that "A careful multi-scalar analysis of the economic and political context in which these gardens act is . . . one of the main tasks of critical and at the same time solidary scholarship" because a better understanding of the social, economic and institutional context in which these initiatives act is necessary "to address these contradictions in (their) daily practices and ultimately transforming the current (neoliberal) logic" (Rosol, 2018, p. 142).

3 See also their respective websites: Retrieved 10 March 2021, from www.futterkreis.de/; https://marktschwaermer.de/de; https://buendnisjungelandwirtschaft.org/; www.oekonauten-eg.de; www.kulturland.de; www.regionalwert-berlin.de/

4 See their respective websites: Retrieved 10 March 2021, from www.restlos-gluecklich.berlin/, https://shoutoutloud.eu/, https://foodsharing.de/, https://querfeld.bio/

5 Retrieved 10 March 2021, from http://tastethewaste.com/info/film

6 More information on their respective websites: Retrieved 21 September 2021, from https://buildingroots.ca/; https://cfccanada.ca/en/Home; https://foodshare.net/; www.mealexchange.com/. For FoodShare, see also (Classens, 2015; Johnston & Baker, 2005)

7 More information on their respective websites: Last accessed 21 September 2021, from www.moissonmontreal.org; www.peoplespotato.com/who-are-we.html; https://restoplateau.com/; www.parole-dexclues.ca. For Resto Plateau see also Rosol (2021).

8 Note that this does not exist in Germany, and there is a lack of data regarding the extent of HFI (Augustin, 2020).

9 Retrieved 28 September 2021, from https://cfccanada.ca/en/About-Us/Mission-and-Vision
10 Retrieved 28 September 2021, from https://cfccanada.ca/en/About-Us/Mission-and-Vision
11 Retrieved 13 October 2021, from https://cfccanada.ca/en/Our-Work/Community-Food-Centres/Call-for-Expressions-of-Interest/Community-Food-Centre-Backgrounder
12 According to map retrieved 19 October 2021, from https://ernaehrungsraete.org/
13 In Berlin, at least some members of the FPC are very much aware of and have always tried to include systemic questions of access and social justice in their demands and actions, also in a global perspective (own observations and interviews 20/2017 and 43/2019 with Berlin FPC members, Berlin). In fact, the initial vision of the Berlin FPC explicitly names justice as one of nine guiding goals and specifies it as follows: "A fair, predominantly regionally structured food system makes it possible to have access to sustainably produced, healthy and culturally appropriate food close to home, regardless of income, education, gender, skin color, cultural background or religion. Our local food system enables global justice by excluding everything that leads here or elsewhere to the exploitation of people or the destruction of their livelihoods. The destruction of the environment is avoided, as is the overexploitation of resources. Producers and workers can systematically rely on living wages/income opportunities – in Berlin, Brandenburg and worldwide" (Ernährungsrat Berlin, 2016, own translation). In the subsequently approved list of 'Demands to the Berlin government', a 28-page-long detailed document, justice or questions of access are not an explicit or prominent demand anymore. The document does contain some reference to social justice, social sustainability, global food justice, the importance of social policy, and demands a social (and environmental) orientation of public subsidies amongst other (Ernährungsrat Berlin, 2017).
14 Educational efforts *can* include enhancing literacy on the – problematic – characteristics of the dominant industrial food system, on the few powerful players that dictate prices to both producers and consumers, on global value chains, and on the reasons why we are in a global food and agriculture crisis, for example (also expressed in Interview 24/2017, Frankfurt), in a collective and empowering sense. However, these efforts seem to be often – or mostly – directed towards educating consumers on how to make better food choices on an individual level, that is, focuses on behavioural, not structural change.
15 This may actually be true in instances, if we look at the price for a certain food item only or when modes of solidarity are enacted (some pay more, others less; the literal translation of the German term for CSA is Solidarity Agriculture – mainly referring to the solidarity between consumer and producers, but in some cases there are also solidary redistribution models among consumer-members) – however, such a view neglects the fact that people in precarious living situations rarely have the resources (time, energy, etc.) to afford the 'little extra effort' or engage actively in groups like a CSA. Active participation in such groups in most cases requires time, frustration tolerance, negotiation, and communication skills. These groups have often rather exclusionary group codes and group dynamics and furthermore often cater to a very specific clientele (white, middle-class, educated, even if perhaps low-income, such as students). In other words: models based on collective organizing and unpaid labour are generally not accessible to everyone. Being able to engage in such organizations already requires a certain amount of privilege (see for a similar observation Gross, 2009) and may also lead to a 'third shift' for women (Som Castellano, 2016; see also Brady et al., 2017).
16 For a pointed critique of such an idea, see Sharzer (2012, p. 7): "Some time ago, I was talking with a nutritionist friend about how expensive and time – consuming it is to be poor. You have to chase low – wage jobs, live in poor – quality housing and endure the daily stress of trying to afford the essentials. Government, which used to provide a social safety net, doesn't help much. Warming to the topic, I added, 'They don't even provide

spaces for community gardens.' My friend replied, 'Why should poor people have to grow their own food?' I had never considered this before. When you're poor, time and energy aren't the only things to go: the first is dignity, as you're forced to scrape by on less. Is there anything noble in adding yet another burden of work?"

17 In Berlin, since August 2019, school lunch is even provided for free up to grade 6 (see www.vernetzungsstelle-berlin.de/projekt-schule/organisation/kosten, last accessed 20 October 2021). In Canada, however, there is no general school lunch system, which not only affects food security but adds to the workload of usually the mothers (who have to prepare lunch boxes and cook a hot dinner at night).

18 Some encouraging newer developments include, for example, that in October 2020 the Berlin FPC organized a public event on access to good food for all, in which food poverty and ways towards increased diversity in the FPC and in CSAs were discussed (https://ernaehrungsrat-berlin.de/oeffentliche-diskussion-zugang-zu-gutem-essen-fuer-alle/). Relatedly, a new initiative "*Alle an einen Tisch*" (All together at the table), started in 2019, explicitly seeks to reach out to refugee and migrant communities. See https://ernaehrungsrat-berlin.de/alle-an-einen-tisch-2/, last accessed 20 September 2021.

References

Alkon, A., & Agyeman, J. (Eds.). (2011). *Cultivating food justice: Race, class, and sustainability*. MIT Press.

Alkon, A. H., & Guthman, J. (Eds.). (2017). *The new food activism: Opposition, cooperation, and collective action*. UC Press.

Allen, P. (2010). Realizing justice in local food systems. *Cambridge Journal of Regions, Economy and Society*, *3*(2), 295–308. https://doi.org/10.1093/cjres/rsq015

Allen, P., FitzSimmons, M., Goodman, M., & Warner, K. (2003). Shifting plates in the agrifood landscape: The tectonics of alternative agrifood initiatives in California. *Journal of Rural Studies*, *19*(1), 61–75. www.sciencedirect.com/science/article/pii/S0743016702000475

Augustin, H. (2020). *Ernährung, Stadt und soziale Ungleichheit: Barrieren und Chancen für den Zugang zu Lebensmitteln in deutschen Städten*. Transcript Verlag.

Beck, A., Haerlin, B., & Richter, L. (2016). *Agriculture at a crossroads: IAASTD findings and recommendations for future farming*. Foundation on Future Farming (Zukunftsstiftung Landwirtschaft). www.globalagriculture.org/fileadmin/files/weltagrarbericht/EnglishBrochure/BrochureIAASTD_en_web_small.pdf

Blay-Palmer, A. (2009). The Canadian pioneer: The genesis of urban food policy in Toronto. *International Planning Studies*, *14*(4), 401–416. http://dx.doi.org/10.1080/13563471003642837

Born, B., & Purcell, M. (2006). Avoiding the local trap: Scale and food systems in planning research. *Journal of Planning Education and Research*, *26*(2), 195–207. https://doi.org/10.1177/0739456X06291389

Brady, J., Power, E., Szabo, M., & Gingras, J. (2017). Still hungry for a feminist food studies. In M. Koç, J. Sumner, & A. Winson (Eds.), *Critical perspectives in food studies* (2nd ed., pp. 81–94). Oxford University Press.

CBC Radio (2020). The current, November 26, 2020: Charities shouldering burden as food insecurity grows. Guests: Valarie Tarasuk, Paul Taylor (online transcript). *CBC Radio*. Retrieved October 14, 2021, from www.cbc.ca/radio/thecurrent/the-current-for-nov-26-2020-1.5817137/thursday-november-26-2020-1.5818344

Classens, M. (2015). Food, space and the city: Theorizing the free spaces of foodshare's good food markets. *Canadian Journal of Urban Research; Winnipeg*, *24*(1), 44–61.

http://search.proquest.com.ezproxy.lib.ucalgary.ca/docview/1769724340/abstract/531C89579CFA4A90PQ/1

Dachner, N., & Tarasuk, V. (2017). Origins and consequences of and response to food insecurity in Canada. In M. Koç, J. Sumner, & A. Winson (Eds.), *Critical perspectives in food studies* (2nd ed., pp. 221–236). Oxford University Press.

De Schutter, O. (2014). *Report of the special rapporteur on the right to food. Final report: The transformative potential of the right to food*. United Nations General Assembly. www.srfood.org/images/stories/pdf/officialreports/20140310_finalreport_en.pdf

Doernberg, A., Horn, P., Zasada, I., & Piorr, A. (2019). Urban food policies in German city regions: An overview of key players and policy instruments. *Food Policy*, 89, 101782. https://doi.org/10.1016/j.foodpol.2019.101782

Ernährungsrat Berlin (2016, März 7). Unsere Vision für eine zukunftsfähige Ernährungs- und Landwirtschaftspolitik in der Region (Vision). *Verabschiedet von der Vollversammlung des Ernährungsrats Berlin*. https://ernaehrungsrat-berlin.de/positionen/vision/

Ernährungsrat Berlin (2017, Oktober 12). Ernährungsdemokratie für Berlin! Wie das Ernährungssystem der Stadt demokratisch und zukunftsfähig relokalisiert werden kann (Forderungskatalog). *Verabschiedet von der Vollversammlung des Ernährungsrats Berlin*. https://ernaehrungsrat-berlin.de/ernaehrungsdemokratie-fuer-berlin/

Friedmann, H. (1993). The policial economy of food: A global crisis. *New Left Review*, *I*(197), 29–57. https://newleftreview.org/issues/i197/articles/harriet-friedmann-the-political-economy-of-food-a-global-crisis

Friedmann, H. (2005). From colonialism to green capitalism: Social movements and emergence of food regimes. In F. H. Buttel & P. McMichael (Eds.), *New directions in the sociology of global development* (pp. 227–264). Emerald Group.

Friedmann, H. (2007). Scaling up: Bringing public institutions and food service corporations into the project for a local, sustainable food system in Ontario. *Agriculture and Human Values*, *24*(3), 389–398. https://link.springer.com/article/10.1007/s10460-006-9040-2

Galt, R. E. (2013). Placing food systems in first world political ecology: A review and research agenda. *Geography Compass*, *7*(9), 637–658. http://dx.doi.org/10.1111/gec3.12070

Goodman, D., DuPuis, E. M., & Goodman, M. K. (2012). *Alternative food networks: Knowledge, practice, and politics* (1st ed.). Routledge.

Goodman, D., DuPuis, E. M., & Goodman, M. K. (2013). Engaging "alternative food networks": Commentaries and research agendas. *International Journal of Sociology of Agriculture and Food*, *20*(3), 425–431. https://doi.org/10.48416/ijsaf.v20i3.184

Gottlieb, R., & Joshi, A. (2013). *Food justice*. MIT Press.

Gross, J. (2009). Capitalism and its discontents: Back-to-the-lander and freegan foodways in rural Oregon. *Food and Foodways*, *17*(2), 57–79. https://doi.org/10.1080/07409710902925797

Guthman, J. (2008a). Bringing good food to others: Investigating the subjects of alternative food practice. *Cultural Geographies*, *15*(4), 431–447. https://doi.org/10.1177/1474474008094315

Guthman, J. (2008b). "If they only knew": Color blindness and universalism in California alternative food institutions. *The Professional Geographer*, *60*(3), 387–397. https://doi.org/10.1080/00330120802013679

Guthman, J. (2014). *Food justice*. University of California. https://critical-sustainabilities.ucsc.edu/food-justice/

Guthman, J., Morris, A. W., & Allen, P. (2006). Squaring farm security and food security in two types of alternative food institutions. *Rural Sociology*, *71*(4), 662–684. https://doi.org/10.1526/003601106781262034

Hamm, M. W., & Bellows, A. C. (2003). Community food security and nutrition educators. *Journal of Nutrition Education and Behavior*, *35*(1), 37–43. https://doi.org/10.1016/S1499-4046(06)60325-4

Hinrichs, C. C. (2000). Embeddedness and local food systems: Notes on two types of direct agricultural market. *Journal of Rural Studies*, *16*(3), 295–303. http://dx.doi.org/10.1016/S0743-0167(99)00063-7

Hoffmann, D., Morrow, O., & Pohl, C. (2019). What's cooking in Berlin's food policy kitchen? *RUAF Urban Agriculture Magazine*, *36*, 37–39. https://research.wur.nl/en/publications/whats-cooking-in-berlins-food-policy-kitchen

Holt-Giménez, E. (2017). *A foodie's guide to capitalism: Understanding the political economy of what we eat*. Monthly Review Press.

Holt-Giménez, E., & Shattuck, A. (2011). Food crises, food regimes and food movements: Rumblings of reform or tides of transformation? *The Journal of Peasant Studies*, *38*(1), 109–144. http://dx.doi.org/10.1080/03066150.2010.538578

Huisken, A., Orr, S. K., & Tarasuk, V. (2017). Adults' food skills and use of gardens are not associated with household food insecurity in Canada. *Canadian Journal of Public Health/Revue canadienne de sante publique*, *107*(6), e526–e532. https://pubmed.ncbi.nlm.nih.gov/28252370

IAASTD (2009). *Agriculture at a crossroads: The global report*. International Assessment of Agricultural Knowledge, Science and Technology for Development. www.fao.org/fileadmin/templates/est/Investment/Agriculture_at_a_Crossroads_Global_Report_IAASTD.pdf

IPCC (2019). *Climate change and land: An IPCC special report on climate change, desertification, land degradation, sustainable land management, food security, and greenhouse gas fluxes in terrestrial ecosystems. Summary for Policymakers*. Intergovernmental Panel on Climate Change. www.ipcc.ch/site/assets/uploads/2019/08/4.-SPM_Approved_Microsite_FINAL.pdf

Johnston, J., & Baker, L. (2005). Eating outside the box: FoodShare's good food box and the challenge of scale. *Agriculture and Human Values*, *22*(3), 313–325. https://link.springer.com/article/10.1007/s10460-005-6048-y

Kepkiewicz, L., Chrobok, M., Whetung, M., Cahuas, M., Gill, J., Walker, S., & Wakefield, S. (2015). Beyond inclusion: Toward an anti-colonial food justice praxis. *Journal of Agriculture, Food Systems, and Community Development*, *5*(4), 99–104. https://doi.org/10.5304/jafscd.2015.054.014

Koç, M., McRae, R., Mougeot, L. J. A., & Welsh, J. (1999). *For hunger proof cities: Sustainable urban food systems*. International Development Research Centre.

Kumnig, S., & Rosol, M. (2021). Commoning land access: Collective purchase and squatting of agricultural lands in Germany and Austria. In A. Exner, S. Kumnig, & S. Hochleithner (Eds.), *Capitalism and the commons: Perspectives on just commons in the era of multiple crises* (pp. 35–49). Routledge.

Levkoe, C. Z. (2003). Widening the approach to food insecurity: The stop community food centre. *Canadian Review of Social Policy*, *52*, 128–132. https://crsp.journals.yorku.ca/index.php/crsp/article/view/33892

Levkoe, C. Z. (2006). Learning democracy through food justice movements. *Agriculture and Human Values*, *23*(1), 89–98. https://doi.org/10.1007/s10460-005-5871-5

Levkoe, C. Z. (2011). Towards a transformative food politics. *Local Environment*, *16*(7), 687–705. https://doi.org/10.1080/13549839.2011.592182

Levkoe, C. Z. (2014). The food movement in Canada: A social movement network perspective. *The Journal of Peasant Studies, 41*(3), 385–403. http://dx.doi.org/10.1080/03066150.2014.910766

Levkoe, C. Z., & Wakefield, S. (2011). The community food centre: Creating space for a just, sustainable, and healthy food system. *Journal of Agriculture, Food Systems, and Community Development, 2*(1), 249–268. https://doi.org/10.5304/jafscd.2011.021.012

Mayer, M., & Boudreau, J.-A. (2012). Social movements in urban politics: Trends in research and practice. In K. Mossberger, S. Clarke, & P. John (Eds.), *The Oxford handbook on urban politics* (pp. 273–291). Oxford University Press.

McIntyre, L., Tougas, D., Rondeau, K., & Mah, C. L. (2015). In-sights about food banks from a critical interpretive synthesis of the academic literature. *Agriculture and Human Values*, 1–17. http://dx.doi.org/10.1007/s10460-015-9674-z

McMichael, P. (2009). A food regime genealogy. *The Journal of Peasant Studies, 36*(1), 139–169. https://doi.org/10.1080/03066150902820354

Mendes, W. (2017). Municipal governance and urban food systems. In M. Koç, J. Sumner, & A. Winson (Eds.), *Critical perspectives in food studies* (2nd ed., pp. 286–304). Oxford University Press.

Moragues-Faus, A., & Morgan, K. (2015). Reframing the foodscape: The emergent world of urban food policy. *Environment and Planning A: Economy and Space, 47*(7), 1558–1573. https://journals.sagepub.com/doi/abs/10.1177/0308518X15595754

Morgan, K. (2015). Nourishing the city: The rise of the urban food question in the global North. *Urban Studies, 52*(8), 1379–1394. https://doi.org/10.1177%2F0042098014534902

Morgan, K., & Santo, R. (2018). The rise of municipal food movements. In S. Skordili & A. Kalfagianni (Eds.), *Localizing global food: Short food supply chains as responses to agri-food system challenges* (1st ed., pp. 27–40). Routledge.

Morrow, O. (2019). Sharing food and risk in Berlin's urban food commons. *Geoforum, 99*, 202–212. https://doi.org/10.1016/j.geoforum.2018.09.003

Myers, J. S., & Sbicca, J. (2015). Bridging good food and good jobs: From secession to confrontation within alternative food movement politics. *Geoforum, 61*, 17–26. http://dx.doi.org/10.1016/j.geoforum.2015.02.003

Pieper, J., Schneider, U., & Schröder, W. (2020). Gegen Armut hilft Geld. Der Paritätische Armutsbericht 2020. *Der Paritätische Gesamtverband*. www.der-paritaetische.de/alle-meldungen/gegen-armut-hilft-geld-der-paritaetische-armutsbericht-2020/

Poppendieck, J. (1998). *Sweet charity? Emergency food and the end of entitlement*. Penguin.

Pothukuchi, K., & Kaufman, J. (1999). Placing the food system on the urban agenda: The role of municipal institutions in food systems planning. *Agriculture and Human Values, 16*(2), 213–224.

Power, E. (1998). Combining social justice and sustainability for food security. In M. Koc, R. MacRae, L. Mougeot, & J. Welsh (Eds.), *For hunger-proof cities* (pp. 30–37). International Development Research Centre.

Power, E., Belyea, S., & Collins, P. (2019). "It's not a food issue; it's an income issue": Using nutritious food basket costing for health equity advocacy. *Canadian Journal of Public Health, 110*(3), 294–302. https://doi.org/10.17269/s41997-019-00185-5

Raphael, D. (2016). *Social determinants of health: Canadian perspectives* (3rd ed.). Canadian Scholars' Press.

Renting, H., Marsden, T. K., & Banks, J. (2003). Understanding alternative food networks: Exploring the role of short food supply chains in rural development. *Environment and Planning A, 35*(3), 393–411. https://doi.org/10.1068%2Fa3510

Riches, G. (2018). *Food bank nations: Poverty, corporate charity and the right to food*. Routledge.
Riches, G., & Silvasti, T. (Eds.). (2014). *First world hunger revisited: Food charity or the right to food?* Palgrave Macmillan.
Rideout, K., Riches, G., Ostry, A., Buckingham, D., & MacRae, R. (2007). Bringing home the right to food in Canada: Challenges and possibilities for achieving food security. *Public Health Nutrition, 10*(6), 566–573. https://doi.org/10.1017/S1368980007246622
Rockström, J., Edenhofer, O., Gaertner, J., & DeClerck, F. (2020). Planet-proofing the global food system. *Nature Food, 1*(1), 3–5. https://doi.org/10.1038/s43016-019-0010-4
Rosol, M. (2010). Public participation in post-Fordist urban green space governance: The case of community gardens in Berlin. *International Journal of Urban and Regional Research, 34*(3), 548–563. http://onlinelibrary.wiley.com/doi/10.1111/j.1468-2427.2010.00968.x/abstract
Rosol, M. (2018). Politics of urban gardening. In K. Ward, A. E. G. Jonas, B. Miller, & D. Wilson (Eds.), *The Routledge handbook on spaces of urban politics* (pp. 134–145). Routledge.
Rosol, M. (2020). On the significance of alternative economic practices: Reconceptualizing alterity in alternative food networks. *Economic Geography*, 96(1), 52–76. https://doi.org/10.1080/00130095.2019.1701430
Rosol, M. (2021). Gut essen in Gemeinschaft. Städtische Ernährungsinitiativen für Begegnung, Gerechtigkeit und Muße. In P. P. Riedl, T. Freytag, & H. W. Hubert (Eds.), *Urbane Muße. Materialitäten, Praktiken, Repräsentationen* (pp. 335–352). Mohr Siebeck.
Rosol, M., & Barbosa Jr, R. (2021). Moving beyond direct marketing with new mediated models: Evolution of or departure from alternative food networks? *Agriculture and Human Values, 38*(4), 1021–1039. https://doi.org/10.1007/s10460-021-10210-4
Rosol, M., & Strüver, A. (2018). (Wirtschafts-)Geographien des Essens: Transformatives Wirtschaften und alternative Ernährungspraktiken. *Zeitschrift für Wirtschaftsgeographie, 62*(3–4), 169–173. https://doi.org/10.1515/zfw-2018-0005
Saul, N., & Curtis, A. (2013). *The stop: How the fight for good food transformed a community and inspired a movement*. Random House.
Scharf, K., Levkoe, C. Z., & Saul, N. (2010). *In every community a place for food: The role of the community food centre in building a local, sustainable and just food system*. Metcalf Food Solutions, The Metcalf Foundation. http://metcalffoundation.com/wp-content/uploads/2011/05/in-every-community.pdf
Schiff, R. (2008). The role of food policy councils in developing sustainable food systems. *Journal of Hunger & Environmental Nutrition, 3*(2–3), 206–228. http://www-tandfonline-com.ezproxy.lib.ucalgary.ca/doi/abs/10.1080/19320240802244017
Sharzer, G. (2012). *No local: Why small-scale alternatives won't change the world*. Zero Books.
Slocum, R. (2007). Whiteness, space and alternative food practice. *Geoforum, 38*(3), 520–533. www.sciencedirect.com/science/article/pii/S0016718506001448
Som Castellano, R. L. (2016). Alternative food networks and the labor of food provisioning: A third shift? *Rural Sociology, 81*(3), 445–469. http://dx.doi.org/10.1111/ruso.12104
Sonnino, R., Tegoni, C. L. S., & De Cunto, A. (2019). The challenge of systemic food change: Insights from cities. *Cities*, 85, 110–116. https://doi.org/10.1016/j.cities.2018.08.008
Spoel, P., & Derkatch, C. (2020). Resilience and self-reliance in Canadian food charter discourse. *Poroi, 15*(1). https://doi.org/10.13008/2151-2957.1298
Swan, E. (2023). White natures, colonial roots, walking tours, and the everyday. In O. Morrow, E. Veen, & S. Wahlen (Eds.), *Community food initiatives: A critical reparative approach*. Routledge.
Tarasuk, V. (2001). A critical examination of community-based responses to household food insecurity in Canada. *Health Education & Behavior, 28*(4), 487–499. https://doi.org/10.1177%2F109019810102800408

Tarasuk, V. (2017). *Implications of a basic income guarantee for household food insecurity*. Research Paper No. 24 Thunder Bay. Northern Policy Institute. www.northernpolicy.ca/upload/documents/publications/research-reports/paper-tarasuk-big-en-17.06.14.pdf

Thurn, V., Oertel, G., & Pohl, C. (2018). *Genial lokal: So kommt die Ernährungswende in Bewegung*. Oekom.

Toronto City Council (2001a). *Toronto's food charter*. https://www.toronto.ca/legdocs/mmis/2018/hl/bgrd/backgroundfile-118057.pdf

Toronto City Council (2001b). *The growing season*. https://sustainontario.com/greenhouse/resource/the-growing-season-food-and-hunger-action-committee-phase-2-report-2001/

UN Economic and Social Council (1999). *Substantive issues arising in the implementation of the international covenant on economic, social, and cultural rights. Comment 12: The right to food*. www.fao.org/fileadmin/templates/righttofood/documents/RTF_publications/EN/General_Comment_12_EN.pdf

Wakefield, S., Fleming, J., Klassen, C., & Skinner, A. (2012). Sweet charity, revisited: Organizational responses to food insecurity in Hamilton and Toronto, Canada. *Critical Social Policy*, *33*(3), 427–450. http://dx.doi.org/10.1177/0261018312458487

Weiler, A. M., Otero, G., & Wittman, H. (2016). Rock stars and bad apples: Moral economies of alternative food networks and precarious farm work regimes. *Antipode*, 48(4), 1140–1162. https://doi.org/10.1111/anti.12221

Wekerle, G. R. (2004). Food justice movements – policy, planning, and networks. *Journal of Planning Education and Research*, *23*(4), 378–386. https://doi.org/10.1177%2F0739456X04264886

Welsh, J., & MacRae, R. (1998). Food citizenship and community food security: Lessons from Toronto, Canada. *Canadian Journal of Development Studies/Revue canadienne d'études du développement*, *19*(4), 237–255. http://dx.doi.org/10.1080/02255189.1998.9669786

Whatmore, S., Stassart, P., & Renting, H. (2003). What's alternative about alternative food networks? *Environment and Planning* A, *35*(3), 389–391. https://doi.org/10.1068%2Fa3621

Winne, M., Joseph, H., & Fisher, A. (1997). *Community food security: A guide to concept, design, and implementation*. Community Food Security Coalition. http://foodsecurity.org/pubs/#cfsguide

Wiskerke, J. S. C. (2009). On places lost and places regained: Reflections on the alternative food geography and sustainable regional development. *International Planning Studies*, *14*(4), 369–387. http://dx.doi.org/10.1080/13563471003642803

Wittman, H., Desmarais, A. A., & Wiebe, N. (Eds.). (2010). *Food sovereignty: Reconnecting food, nature and community*. Fernwood Publishing and FoodFirst Books.

Ziegler, J. (2008). *Promotion and protection of all human rights, civil, political, economic, social, and cultural rights, including the right to development*. Report of the Special Rapporteur on the right to food, Jean Ziegler. UN General Assembly. Human Rights Council. www.righttofood.org/wp-content/uploads/2012/09/AHRC75.pdf

Part 2

Cooperatives, cooperation, and concerns in CFIs

6 Constraint and autonomy in the Swiss "local contract farming" movement

Jérémie Forney, Julien Vuilleumier, and Marion Fresia

In the Swiss context, local contract farming (LCF)[1] refers to collectives in the form of associations of producers and consumers connected by a long-term contract and by their shared goal of supporting both local agriculture that respects the environment and fair pay for the farmers.[2] In this system, the consumers, by paying a subscription fee, commit to supporting the production of the growers. In return, they receive, either monthly or weekly, a "basket" of food products whose quantity, quality, and price have been established in advance. These collectives of producers and consumers, which are part of the legacy of the pioneering experiences of the 1970s and 1980s (the *Jardins de Cocagne* in Geneva and the *Clef des Champs* in the Jura region) that were driven by a concern for the environment and a desire to challenge consumerist society, have enjoyed a revival since the 2000s. Food scandals, health concerns, the deregulation of the agricultural market, and the mainstreaming and industrialisation of organic food have all contributed to a (renewed) appreciation for local supply chains, which are perceived as being safer, more ecologically sound, and in the spirit of greater solidarity. Thus, the number of LCF ventures has exploded in recent years in French-speaking Switzerland, growing from 13 in 2003 (Porcher, 2010) to 28 in 2016, with a total membership of approximately 6300. Furthermore, according to the 2016 URGENCI report (Volz et al., 2016), 60 such ventures have been counted in Switzerland as a whole. In a sign of the growing institutionalisation of such collectives, a Federation of French-speaking Swiss Local Contract Farming (*Fédération Romande d'Agriculture contractuelle de proximité, FRACP*) was formed in 2007. Their organisational models have also begun to diversify: alongside the traditional model of the first ventures, which directly integrate production into a cooperative or associational structure, another model has developed, of the association as an intermediary, linking consumers to producers who are not employed by the LCF but who remain legally and economically autonomous.[3] Thus, LCF systems can no longer be seen as a marginal phenomenon in Switzerland. Not only that, but all over the world, similar arrangements, though under a variety of names and with important differences in emphasis, are multiplying and networking with each other: whether it is the Associations to Support Small-Scale Farming (AMAP) in France, Community-Supported Agriculture (CSA) in the United States and Canada, Solidarity Agriculture (Solawi) in Germany, or the *teikei* system in Japan. Through their

DOI: 10.4324/9781003195085-8

existence and actions, these solidarity purchasing initiatives have become the contemporary face and symbol of healthy consumption that respects both ecosystems and food producers.

If these food networks are the embodiment of new forms of agri-food utopias, then, they are not utopias in the sense of a non-existent "elsewhere" but rather in the form of a set of discourses, intentions, and concrete experimentations (Stock et al., 2015). Their utopian dimension can be found in the "promise of difference" that they express, in other words the "promise of a different way of organizing food production, trade, and/or consumption and the promise of the benefits associated with that" (Le Velly, 2017, p. 24). In other words, by transforming the relations and interactions between producers and consumers and developing new visions about them, these food networks hope to create better – understood as fairer and greener – food systems

In this chapter, we propose to take this promise seriously, in other words to "recognize the intention without assuming that the operating methods are radically different" (Le Velly, 2017: 24) and to examine the concrete changes brought about by ventures connected to LCF. Starting with the case of Switzerland, we will attempt to answer two large questions. First, how is this promise embodied in a variety of experiences in French-speaking Switzerland and within specific reorganisations of the producer-consumer relationship? And second, what are its effects on the practices of production and consumption and on the subjectivities of the actors participating in those experiences? We will focus on two elements that constitute essential vehicles for the rearrangements that we have observed: the contract that ties members to the initiative and the food itself. The contract translates and formalises the promise of a better food system by reinforcing the producers' autonomy through a long-term commitment on the part of the consumers. The food, whose quality and value are redefined, acts as a mediator in the redefinition of the producer-consumer relationship. By means of these empirical investigations we will attempt to move beyond the debate about whether these networks are truly alternative or not, in order to look at these ventures as particular collections of human and non-human actors that "open up spaces in which to enact a politics of possibility" (Harris, 2009, p. 58). We will pay particular attention to the process of empowerment for the members – producers and consumers – that is brought about by participation in LCF.

Like Le Velly (2017), we believe that an examination of the transformations brought about by LCF must remain central to the analysis, but also that we need to leave behind a framework that sees these changes only in terms of an "alternative" model to conventional agriculture or, on the other hand, as a simple "niche market" that can then be absorbed into conventional agriculture. We take as our starting point the observation that LCF is not positioned "against" or "outside of" the dominant agri-food systems but that it maintains, instead, a dialectical relationship of differentiation and integration with respect to those dominant systems. The vision of an alternative model that is completely outside the industrialised and globalised agri-food system, just like the vision of a neoliberal economic model that could absorb and neutralise all subversive ventures, cannot withstand empirical analysis (Forney & Haeberli, 2017; Chiffoleau et al., 2019; Matacena & Corvo, 2020). Such a vision does not allow us to grasp the multiplicity of moralities, institutional arrangements, and economic rationales that are threaded through contemporary agri-food systems and, more broadly, all the various institutions of the market economy. Thus, while recognising how these so-called alternative networks are in fact woven into the globalised capitalist economy, we will make the interpretive choice of "reading for difference rather than dominance" (Gibson-Graham, 2008). From this perspective, it seems relevant to refer to the literature on the concepts of autonomy and empowerment in the agricultural and food systems. These concepts, which are central to the literature on family farming (Mooney, 1988; Stock & Forney, 2014), on small-farm movements (van der Ploeg, 2008), on food sovereignty movements (Trauger, 2015), and on cooperative practices in agriculture (Stock et al., 2014; Lucas & Gasselin, 2018), are largely absent, surprisingly, from the literature on CSA and similar initiatives, despite an, arguably, shared objective of getting free from some of the dependencies related to the dominant food system. Wilson (2013), inspired by the work of Gibson-Graham (2006) and Chatterton (2005), is an exception, proposing the concept of "autonomous food spaces." In this approach, autonomy is understood both as a process of self-assertion (subjectivity) and, at the same time, as a process of reorganisation of social, economic, and political relations, the outcome of which is always partial and uncertain. Another discussion around notions of autonomy and autonomisation processes contrasts individualistic with collective understandings of autonomy. Stock et al. (2014) deconstructs neoliberal assumptions of autonomy as individual freedom, emphasising the collective autonomisation processes through cooperation. In a similar vein, Emery (2015) analyses how individualistic representations of independency have been ideologically imposed on English agriculture, at the expense of collective engagements that would have resulted in "actual independence." These discussions on "autonomous food spaces" and collective processes of autonomisation seemed to us to be particularly pertinent as we analysed our data. We will see, in fact, how they allow us to conceive of these ventures simultaneously as an attempt to build agri-food systems that can at least partially liberate their participants from the constraints of the dominant organisation of the market and, at the same time, as a way to promote critical reflection and a willingness to engage in transformation in their members.

The varying models and trajectories of three ventures

Our study is based on a sustained ethnographic study of three associations over the course of almost four years (2012 to 2016), including an analysis of their official documentation and field notes from participant observation at their meetings, work in the fields, or the distribution of vegetable baskets; 20 semi-directed interviews with committee members, producers, and consumers; and, finally, a longitudinal study of 15 consumers over the course of 18 to 24 months including an initial interview, a self-evaluation by way of a food budget, and a final interview based on the completed budget. This diversity of data has been analysed as single corpus of ethnographic material. The coding has been inductive, following the principles of grounded theory (Glaser & Strauss, 2006) to create analytical categories from the ethnographic material itself. All the interviews and observations were conducted in French. The quotes reproduced in this chapter have been translated into English by the authors.

In this section, we present our three case studies with the intention of showing the variety of histories, commitments, and trajectories that are included in local contract farming in French-speaking Switzerland and the promise of difference that it embodies.

Rage de Vert: *gardening to create connections and debate*

"We wanted to change our lives and garden, just plain garden and feed people, it was a very simple thing."[4] This is how one of the two founders of this venture describes the beginnings of *Rage de Vert* ('Fury for Green'). He and his co-founder had to start from scratch, throwing themselves into organic market gardening after a brief experience on a farm in the region. Our interviewee is a biologist by training and his teammate a photographer; together, they enlisted their friends and acquaintances to found the *Rage de Vert* association. In the tradition of the pioneering experiences of the 1970s and 1980s, they embodied the paradigm of the new farmer who takes up farming with no direct family connections to farming nor any training in the field to begin with (Rouvière, 2015). Faced with the challenge of finding land to cultivate, *Rage de Vert* was able to begin its work in 2010 when the city of Neuchâtel provided it with urban brownfields (abandoned post-industrial land). In 2011, the association delivered its first baskets to about a hundred members. The basics of the operation were already in place at that point: it was heavily urban; it relied on volunteer work by its members; the weekly distribution allowed for a moment of connection between the gardeners and the members; and collaborations with other actors began to take place, making it possible to obtain an additional plot on which to plant winter vegetables. After its hesitant beginnings, *Rage de Vert* became increasingly professionalised: the committee took a stronger role and, beginning in 2015, the system was improved with the introduction of online tools for managing members, deliveries, and volunteer work. In 2014, the hiring of a new gardener also made it possible to develop educational activities, with the introduction of specific modules for schools. In this consolidation, the number of members grew from approximately 100 in 2011 to almost 180 in 2017, following some serious promotional activity. Since its inception, *Rage de Vert* had always faced some uncertainty about the land on which it farmed, given that the two borrowed plots could be taken back from one year to the next. In 2015, the association learned that one of the plots was going to be developed and that they urgently needed to find an alternative. They found some garden market plots that they were able to rent long-term from a foundation involved in socio-vocational integration. In 2016, after a lively debate, the committee voted in favour of the move: the new plots were more than ten kilometres from the city, which called into question the organisation of educational activities and the urban nature of the venture.

The potential for "difference" and transformation towards more sustainability expressed by *Rage de Vert* is rooted in what it calls inclusive production that is respectful of the environment and that also seeks to promote the use of urban brownfields, "soft mobility" (non-motorised transport), and social bonds. Although it was not entirely able to maintain its urban dimension, the association continues to cultivate its social ties with a variety of audiences (in particular, in the context of its participation in programmes to integrate asylum seekers or the long-term unemployed). The gardeners are also in contact with the consumer members every week when the baskets are distributed, as well as during farm work, in which the consumers are required to participate for two half days every year.

Les Jardins d'Ouchy: *supporting farms on a human scale*

In 2007, in Lausanne, the association *Les Jardins du Flon* was launched; the first baskets were delivered at the end of the summer. Supported by leftist politicians, this venture was started by urban consumers who allied themselves with three regional producers. Soon, *Les Jardins du Flon* was fully subscribed, and due to strong demand from consumers, a second association was founded in another district of Lausanne: *Les Jardins d'Ouchy*. Taking up the model of *Les Jardins du Flon*, which it adapted, *Les Jardins d'Ouchy* was based above all on the principles of support for family farming and proximity between producers and consumers. Production followed the norms of integrated production established by the federal agricultural policy of Required Ecological Services (*Prestations Ecologiques Requises*, PER),[5] but it was not required to follow the norms of the "bio suisse" label. *Les Jardins d'Ouchy*'s two farmers are based in the greater Lausanne area. Only a portion of their vegetable production goes to the LCF baskets; they are involved in other farming ventures (notably fruit production for wholesalers) and have other sales outlets, particularly for mass retail. Starting with the first delivery, in September 2009, the system was established very simply: the two farmers coordinate production and delivery and define the contents of the baskets. They take turns being present for the distribution, which takes place every Thursday in a house in the neighbourhood.

Les Jardins d'Ouchy grew very quickly, starting with an initial membership of almost 140 households and reaching almost 200 after a year and a half. Beginning in 2013, its membership shrunk, with the number reaching about 170 in 2017. One of the primary concerns of the coordinator is to get the consumer members to participate in the association's activities beyond simply paying their annual subscription fee. Although their active participation is limited, supporting "local farmers" emerges as a key motivation for the consumers. In its vision statement, the association also emphasises its desire to "preserve the farming profession by making it possible to maintain human-scale farms." This echoes the political and unionist dynamic that continues to be strongly present in official speeches by the founders of *Les Jardins d'Ouchy*. The issues of preserving the environment or promoting farming that is 100% organic, or raising the awareness of a broader public on these issues, along with creating a sense of belonging to a collective, appear to be secondary here.

Notre Panier Bio: ***promoting organic production and food sovereignty***

Organic and certified above all is the watchword for the farmers of *Notre Panier Bio* (Our Organic Basket) in the canton of Fribourg. The group that is behind this venture, which was created at the end of 2006, had its origins in the association of organic farmers of Fribourg and interested consumers and was directed by an agricultural engineer who became the secretary of the group. An organic market gardener also played a key role, in particular by taking care of the group's logistics and administration, from the very beginning, through his own business. The first baskets were delivered at the beginning of 2007, with 11 associated producers and about 50 members. The distinguishing feature of *Notre Panier Bio* is to offer baskets at a variety of distribution points (about 40 of them in 2017) around the canton, but also to bring together a large number of producers, 24 in 2017. As the current coordinator puts it, "We want to make organic food accessible to a large number, to people who do not have access to organic grocery stores or to markets, and at the same time, we offer small producers a chance to become widely known, which they would not otherwise have."[6] This model, which includes the delivery of monthly baskets and a large variety of products (including processed food, dairy, grains, and meat), complements the direct sales that the vast majority of the producers practice in addition. Thus, none of the producers sells the majority of their production through *Notre Panier Bio*. The value of the network of producers involved has made it possible for the association to grow rapidly, from 400 members a year after it was founded to almost 640 households in 2018. However, this number has decreased again in the following years. In December 2021, there were only 427 contracts signed with consumer-members.[7] Indeed, the association still has important work to do in building the relationship between its producers and its members, a relationship which at this point is still more imaginary than it is embodied in any physical encounters. This situation is due to the distribution structure that does not bring together producers and consumers, and, probably, to the relatively large size of the association which dilutes interpersonal relationships. The association is also constantly working to produce the "quality" of its products, which it defines in terms of ecological and taste criteria. Thus, the promise of difference expressed by *Notre Panier Bio* lies in its offering of organic products, which is an argument that is constantly emphasised in this group. But the aspects of solidarity with Swiss producers and of food sovereignty are also present, as evidenced by an extract of the group's charter: "With solidarity and fairness, the contractual system reinforces the independence of the farms and promotes direct sales and consumption."

The diversification of commitments; interdependencies

Our three case studies belong to the "second wave" of producer-consumer collectives that began to appear in the early 2000s, in the context of a new appreciation for short supply chains against the backdrop of health, social, and ecological crises. They attest to the growing multiplicity and diversity of models and kinds of commitment to be found in these collectives.

Compared to the traditional model of a production unit that is directly embedded within a cooperative or associative structure (exemplified here by *Rage de Vert*) we note the diversification – confirmed among the members of the regional federation, the FRACP – into models where the association is an intermediary (see *Les Jardins d'Ouchy, Notre Panier Bio*). In these new models, the approach appears to be a more pragmatic one than what we see among the pioneers of the movement, although the critique of the industrial agri-food model remains the same. This pragmatism is characterised by a high degree of intertwinement with more conventional sectors, in particular for *Notre Panier Bio* and *Les Jardins d'Ouchy*. In both cases, the producers are involved in a variety of sectors, ranging from direct sales to delivery to wholesalers and collaboration with the dominant players in mass retail. The producers only deliver a small part of their production to the LCF groups, though they receive a great deal of recognition for it. Thus, there is a high degree of "hybridization" (Filippini et al., 2016) between the so-called conventional sectors and the so-called alternative ones. The consumer members, similarly, practice multimodal provisioning, of which mass retail remains an integral part while LCF only provides a modest portion of their food. This is true for *Rage de Vert* as well. These findings show that the boundaries between the so-called alternative and conventional spheres are largely permeable, and that the former continue to be at least partially dependent on the latter. This statement echoes findings in diverse economies" research (Wilson, 2013; Forney & Häberli, 2017; Blumberg et al., 2020), inspired by the seminal work by Gibson-Graham (1996). In their perspectives, looking for the "cracks" in the capitalist hegemony allows researchers to overcome the capitalocentric framing at the basis of the binary alternative-conventional. It encourages us to switch the focus from food "alternatives" to food "diversity" (Cameron & Wright, 2014, p. 2). By nuancing the domination of capitalist logics, highlighting the possibility of diverse economic models, and nuancing the dichotomy between alternative and conventional, such approaches reintroduce room for hope, without having to naively ignore the troubles that initiatives in the like of LCF encounter in their enactment of different values in food systems.

One last observation that follows from our three case studies is the relatively homogeneous sociological profile of the consumer members. While the LCF models have multiplied and diversified over time, they have still not managed to achieve diversity in their membership, which is comprised mostly of people with an above-average education level (though not necessarily above-average income). Thus, at *Rage de Vert*, almost 80% of the membership has a tertiary education (university or higher education), whereas for the Swiss population as a whole that percentage was 41% in 2016[8]. In terms of age, they are mostly in their 30s (54% are between 30 and 39 years old), and the income of the member households matches the national average. The members of *Notre Panier Bio* present a similar profile. In addition, the initiators and some of the committee members of the three ventures also share a relatively similar socio-demographic profile. This finding corroborates what has been observed in other countries (Paranthoën, 2013; Jaffe & Gertler, 2006), suggesting that these collectives are, at present, primarily a "class" phenomenon (Brusadelli et al., 2016).

Table 6.1 Comparison of the three case studies[9]

	Rage de Vert	*Notre Panier Bio*	*Les Jardins d'Ouchy*
Mode of organisation	Non-profit association[10]	Non-profit association	Non-profit association
Starting year	2010	2006	2009
Producers	4 "gardeners" employed by the association (part-time)	24 organic farmers, all independent businesses	2 main farmers, 4 occasional farmers, all independent businesses
Number of members	180	640	170
Products	Vegetables	Vegetables, fruits, legumes, dairy, cereal, eggs, meat	Vegetables, fruits
Standard	Organic (Bio Suisse)	Organic (Bio Suisse)	Conventional (Swiss federal standard)

At the core, the contract and a long-term commitment by consumers

Beyond the diversity of institutional models followed by the second-wave LCF systems, all of them define and create the producer-consumer relationship by means of a long-term contract, which commits the consumer to one or more producers for at least a year. In LCF, the contract is an important tool to steer and enact the collaboration between producers and consumers according to the values promoted by the initiatives.

The contract defines the terms of the engagement according to specific characteristics that diverge from the most common forms of food provisioning. These are generally:

- a long-term commitment by the consumers, for 12 months in the cases studied here, with a prepayment of the annual amount of the transaction.
- following certain (more or less strict) environmental production norms and fair pay for the producers.
- a collective negotiation of prices, proposed by the producers and then agreed on at the general meetings.
- an obligation on the part of both the producers and the consumers to become members of the association and to pay a membership fee.
- and finally, in some cases, participation by the consumers in the agricultural work for a few days per year (sweat equity) (as in the case of *Rage de Vert*, where the consumer members have to contribute two half days of volunteer work per year).

At first glance, the contract would appear to clearly favour the empowerment of the producers. Indeed, the producers do gain economic security, by receiving prices that they consider fair and by sharing a certain amount of risk with the consumers. In addition, the length of the commitment allows the farmers to make a specific plan for their crops and harvests over the course of a season, and the prepayment gives them a certain level of security, both for their initial investments and also in the face of weather-related and other risks that might affect production.

Thus, the LCF system facilitates the startup of a farming operation for new farmers by guaranteeing them a minimum sale for their production, while for farmers who are already established, it allows them to diversify their outlets and become at least partially independent of the large retail chains. In addition, our interviews make it clear that through their participation in LCF, the producers receive renewed appreciation for their profession and a greater recognition of its arduousness and socioeconomic challenges. The importance of recognition is explicit in the work of this farmer:

> If we didn't have the dialogue with the members when we participate to the distribution [of the baskets], it wouldn't be worth the trouble. You know, they ask you questions that looks a bit basic. But this is so important to provide answers and explain what we do, how we do it. Then, they say thank you for having given something to them. That's really what motivated me when I say to myself that there a lot of work in preparing all this market [i.e. the site of distribution] and coming here, to town.
>
> (C., producer for Jardins d'Ouchy, 12.06 2016, Lausanne)

Other types of direct interaction with consumers, for instance at the farmer market, can also attract the same compliments on products and related pride. However, the contract creates a stronger feeling of belonging to a collective. In addition, these networks also offer the potential for greater political commitment on behalf of small farming, which the producers very much appreciate. And finally, the fact that the consumers appreciate and comment on the quality of their products gives the farmers great satisfaction. In our interviews, the contacts with the consumers, even if they are only intermittent, are a fundamental element of the farmers' motivation for participating.

At the same time, however, this connection with the consumers and the participation in LCF also involve new obligations for the producers, such as delivery and distribution, as well as participating in an associational structure that entails the development of new skills in association management, volunteer management, the organisation of events and farm visits, etc. All of this takes time and commitment, which can weigh on the producers in the long term, especially producers for whom LCF is not the only marketing channel.

The advantages of the contract in terms of providing support for the producers, mentioned earlier, are made possible, in part, by the transfer of certain constraints to the consumers. For the consumer, belonging to an LCF association means giving up, at least in part and mostly for vegetables, the greater freedom of choice and price that they enjoy in the context of mass retail shopping. This kind of long-term commitment, connected with prepayment for the goods they receive, involves a certain economic effort. In addition, the contents of the basket the consumers receive is predefined and imposed on them: the consumer has to adapt to the offerings of one or a few producers, which then vary according to the season, the weather conditions, or other hazards. The rhythms of delivery, whether weekly or monthly, as well as the quantity are also settled in advance, and collectively, by the contract.

And yet, the consumer members that we interviewed actually most often put a positive value on these constraints: they were presented as being both chosen and embraced, even valued. While the consumers do have to give up a certain form of individual liberty, they say that they do that in order to better fulfil their commitment to patterns of consumption that are directed, above all, towards a desire to preserve the environment, to support Swiss farmers, or towards a quest for products that are healthy or that present new tastes. These two quotes illustrates the diverse valuations, if not of the constraints themselves, of their effects:

> One thing I did not mention and that's very important . . . it's the solidarity with farmers in prices. It's sharing risks and chances they have regarding weather . . . and all the hazards of nature. And this, I find really beautiful to pass on to our children. Myself too. It makes me feel better.
>
> (with S., member of Jardin d'Ouchy, 6 January 2016, Lausanne)

> Well, I discovered that in winter it's all roots. I discovered veggies that I had never cooked before, tastes that I never cooked before. So, that was interesting to find veggies that I didn't know and that I would never have bought myself.
>
> (F., member of Notre Panier Bio, 5 May 2015, Marly)

The contract, thus, allows them to reaffirm these choices and, above all, to uphold them over time, without having to repeat the effort of (re-)commitment with every act of consumption. Given the complexity of the stakes for sustainability that are involved in the question of food, binding together transportation, agricultural production methods, transformation and conservation, and north-south relations, the act of consumption requires a whole series of translations and processes of delegating choice. To commit to an LCF group, then, is to make a choice, for a relatively long period of time, to radically delegate these negotiations to a small local collective. As certain members explain it, this simplifies the commitment:

> In the baskets, there is what there is, that's it, period. It puts us a little behind what our predecessors did: nature gives you what it gives you and you work with that, you limit your choices, you make do with what you have.
>
> (E., member of Rage de Vert, 21 November 2015, Neuchâtel)

However, it is also because these ventures are not in fact the only source for "food links" in which the consumers are involved that they promote an overall positive experience. In addition, some members do not honour their commitment over the long term. They end up feeling a certain weariness with the lack of diversity in the products on offer (especially in the winter) or with the lack of flexibility in the distribution methods. Thus, there is a turnover of 10% to 15% within each of the three ventures.

Food at the heart of the promise and transformation

LCF initiatives make a "promise of difference," a promise to construct a way of producing, exchanging, and consuming food that avoids the failures of the dominant food system in terms of social and environmental sustainability. This promise has to do primarily with food itself, in its various qualities, which are simultaneously physical, nutritional, taste-related, and social. On the next level, the promise connects all of the social and environmental relations involved in the production, distribution, and consumption of food. We can hypothesise that if consumers – or in this case, more precisely, eaters – are willing to invest differently and to accept certain unaccustomed constraints, it is above all because of this plural nature of food, which crystallises these various relationships between people, and with the natural elements that are part of the process of food production. Food stuff make these connections concrete and tangible. The materiality of food contributes directly to making the consumers willing to develop personal investments in terms of economics (the price), time (for preparation and consumption), and the acquisition of new skills.

Becoming involved with an LCF collective also makes it possible to connect a certain critical approach towards consumption with the acquisition of new practical skills and knowledge about food, connected with the need to adapt to the rhythm of the seasons and sometimes to cook food items or varieties of vegetable with which one was previously unacquainted. This development is connected to the constraints that are built into LCF, such as limited choice and a certain monotony of the supply:

> Yes, well I discovered that winter means root vegetables. I discovered vegetables that I had never cooked, tastes that I had never cooked, and so it was interesting to find vegetables that I didn't know, that I would not have bought.
>
> (F., member of Notre Panier Bio, 8 May 2015, Marly)

These skills and knowledge circulate and are shared within the collective that is made up of the producers, the consumers, and the association itself. As an example, the three ventures all provide their members with recipes or with information on food preservation. This is a way to support the consumers' learning process, their discovery of new tastes, a different aesthetic, or new information about how food is produced. This aspect is particularly appreciated for its educational value within the members' families:

> What I liked about this setup was being able to show my kids, who were both small at the time, how food is grown. Because I myself don't have a house or a garden, and it seemed important to pass on this knowledge to them. So essentially, it was more for them, and for us, that was more secondary.
>
> (E., member of Rage de Vert, 21 November 2015, Neuchâtel)

At the heart of this knowledge is the food, which is much more than just an abstraction: every vegetable calls for its own questions, and the answers are always rooted in specific practices. This learning dynamic can also generate other, more political, forms of engagement. The collectives that we studied, for example, have mobilised to advocate for the inclusion of the concept of food sovereignty in Swiss law (Vuilleumier, 2017). This example shows that a political reflection that is incorporated into consumer practices on an individual level can shift, in certain cases, towards more classic and institutionalised forms of political mobilisation and towards new expressions of "food activism" (Counihan & Siniscalchi, 2013).

Conclusion: moving towards empowerment through collective commitment

By rooting the relationships between producers and consumers in a long-term contract geared towards supporting human-scale farms and modes of production that seek to respect the environment and seasonality, LCF networks are raising anew the question of autonomy in the eating habits of their members. For the producer members, their involvement in the collective allows them to begin or expand on a process of diversification and emancipation from certain dependencies connected with how the dominant markets operate, in particular with respect to how prices are set and how product quality is defined. Of course, most of these aspects are negotiated within the associations, and the producer still has to submit to the collective's organisational principles. Nevertheless, the power relationships in the collective are of an entirely different nature than they are in the mass retail market, and they are much more favourable for the producers. The feeling of proximity and a fundamental willingness to support farmers lead to fairer relations. The empowerment of the producer is thus at the heart of the LCF project.

On the other hand, what the consumer members agree to could appear to represent a loss of autonomy in comparison with the dominant modes of consumption, at least if we reduce autonomy to the aspect of freedom of individual choice. This resonates with the abundance of products available in the wealthy Swiss retail sector. Consumers have certainly no individual control over the offer in supermarkets. However, the latter provide a huge diversity of products, notably in terms of quality, provenance, and price. Local organic products are sitting next to imported cheap food, allowing very diverse modes of consumptions within the boundaries of this assortment. Nevertheless, this individualistic conception of autonomy is contradicted by both the discourses and the practices connected with LCF. While these ventures seek to liberate themselves, at least to some degree, from the large retail chains, they do not do it by setting up an individualistic independence as its alternative. On the contrary, they establish forms of collective empowerment that take the path of forms of chosen interdependence within restricted collectives. In other words, LCF is reframing autonomy as interdependence. This approach towards autonomy sends us back to recent works on the processes of empowerment linked to cooperative practices in more traditional agricultural contexts (Stock et al., 2014; Emery, 2015; Forney & Häberli, 2016, 2017). These texts propose a definition of autonomy that includes collaboration and interdependence, within a relatively symmetric system of power relations. Thus, in their accounts, the consumer members of the LCF ventures emphasise benefits that we can characterise as a gain in autonomy when compared to the constraints imposed by the retail chains: a kind of consumption that is closer to the ideals of sustainability and solidarity that they seek, nourished by a critical and political reflection. Thus, paradoxically, it is the consensual renunciation of certain freedoms of consumer choice and the acceptance of strong interdependencies that produces a feeling of empowerment and autonomy. Admittedly, this renunciation is made easier by the structural context of abundance, which allows the actors to continue enjoying multimodal consumption elsewhere, and to leave an LCF venture at any time. Thus, we have seen that like other so-called ethical consumer practices (Lamine, 2008), LCF ventures only provide a modest portion of their food to consumers, whose provisioning continues to be enmeshed with the large retail chains; in addition, in two of our case studies, the producers also continue to be partially bound into the mass retail system to sell their goods. In addition, these ventures reveal clear limitations, in particular regarding their ability to reach a greater diversity of social classes and to generalise. Our results therefore confirm some of the limitations and troubles characterising alternative food networks and that have been already described in the literature (e.g. Tregear, 2011). Nevertheless, these important nuances do not invalidate the argument of empowerment. Participant to LCF initiative get some level of autonomisation, and are allowed to develop collectively a hopeful engagement for the transformation of food systems. The limitations and troubles met in the process only emphasise the partial nature of this empowerment, which is not built within a completely antagonistic relationship between two totalities – the alternative and the conventional – but rather in the intermingling and coevolution of a variety of economic practices, sometimes contradictory but always connected. Troubles and hope coexist in these attempts to create spaces of autonomy.

Alternative food networks have sometimes been critically described as relying on a neoliberal concept of governance, based on the individual responsibility of the consumer rather than public intervention (Guthman, 2008). However, in fundamentally redefining the modalities of the act of consumption, particularly through long-term subscription and by restricting product choice, the subject that is constructed by the LCF system departs from the principles of individualisation and free choice that are generally associated with neoliberalism. Thus, the constitution of a collective around commitment to LCF itself creates new sites for a collective commitment, which have the distinctive feature of bringing producers and consumers together within a food association landscape that is very segmented.

Therefore, what is produced by participation in LCF ventures is a process of experimentation using concrete proposals oriented towards other potential food futures. Likewise, the LCF systems are part of maintaining, consolidating, and even developing diversified production and distribution channels while at the same time constituting potential sites of a broader mobilisation and contestation of the agri-food industry. The LCF movement can thus be seen as a "food utopia," in the words of Stock et al. (2015). Far from being a completed experience, it expresses an intention and experiments with new relationships to food and new ways of arranging the relationship between producers and consumers by constituting spaces of partial food empowerment. It embodies a "hope," which admittedly only affects certain social groups, but which is nevertheless expanding globally, creating networks, and already affecting a large number of countries, well beyond Europe, North America, and Japan (Shi et al., 2011).

Notes

1 The term "local contract farming" was formalised and brought into widespread use with the establishment of the Federation of French-speaking Swiss Local Contract Farming in 2007. The federation created a charter to define this system and the basic principles shared by the various ventures that were already in existence and already very diverse.
2 As detailed in the chapter, LCF initiatives can take many forms, and food producers are not always "farmers" with a proper farm. In some instances, the food producers are gardeners, employed by the LCF initiatives. In this chapter, however, we use "farmer" as a general term to speak of the people professionally active in agricultural production. This use of the term is also consistent with that of LCF members.
3 This research took place in the context of the Swiss National Science Foundation's research project on "Healthy food and sustainable food production in Switzerland" (PNR 69).
4 Interview with T., founder of *Rage de Vert*, Neuchâtel, 24 April 2014. Unless otherwise indicated, the quotations in this paragraph are also from this interview.
5 The PER norms represent the environmental standard set by federal agricultural policy, which is the condition for access to the direct payment system.
6 From a discussion with M., the coordinator of *Notre Panier Bio*, following a committee meeting, 6 June 2014, Grangeneuve.
7 This decrease in members is probably related to the development of several other LCF initiatives in the same geographical area. Arguably, this boom created some sort of competition among initiatives. However, on the basis of our data, we cannot confirm neither contradict this hypothesis.
8 See the website of the Swiss Federal statistical Office, https://www.bfs.admin.ch/bfs/en/home/statistics/education-science/level-education.assetdetail.22024466.html [consulted the 23.02.2023].
9 The data in the table reflect the situation as in 2018.
10 According to Article 60 of the Swiss Civil Code.

References

Blumberg, R., Leitner, H., & Cadieux, K. V. (2020). For food space: Theorizing alternative food networks beyond alterity. *Journal of Political Ecology*, *27*(1), 1–22.

Brusadelli, N., Lemay, M., & Martell, Y. (2016). L'espace contemporain des "alternatives." *Savoir/Agir*, *4*, 13–20.

Cameron, J., & Wright, S. (2014). Researching diverse food initiatives: From backyard and community gardens to international markets. *Local Environment*, *19*(1), 1–9.

Chatterton, P. (2005). Making autonomous geographies: Argentina's popular uprising and the "movimiento de trabajadores desocupados" (unemployed workers movement). *Geoforum*, *36*, 545–561.

Chiffoleau, Y., Millet-Amrani, S., Rossi, A., Rivera-Ferre, M. G., & Merino, P. L. (2019). The participatory construction of new economic models in short food supply chains. *Journal of Rural Studies*, 68, 182–190.

Counihan, C., & Siniscalchi, V. (2013). *Food activism: Agency, democracy and economy*. Bloomsbury Publishing.

Emery, S. B. (2015). Independence and individualism: Conflated values in farmer cooperation? *Agriculture and Human Values*, *32*(1), 47–61.

Filippini, R., Marraccini, E., Houdart, M., Bonari, E., & Lardon, S. (2016). Food production for the city: Hybridization of farmers' strategies between alternative and conventional food chains. *Agroecology and Sustainable Food Systems*, *40*.

Forney, J., & Häberli, I. (2016). Introducing "seeds of change" into the food system? Localisation strategies in the Swiss dairy industry. *Sociologia Ruralis*, *56*(2), 135–156.

Forney, J., & Häberli, I. (2017). Co-operative values beyond hybridity: The case of farmers' organisations in the Swiss dairy sector. *Journal of Rural Studies*, *53*, 236–246.

Gibson-Graham, J. K. (1996). *The end of capitalism (as we knew it): A feminist critique of political economy*. University of Minnesota Press.

Gibson-Graham, J. K. (2006). *A post capitalist politics*. University of Minnesota Press.

Gibson-Graham, J. K. (2008). Diverse economies: Performative practices for "other worlds." *Progress in Human Geography*, *32*(5), 613–632.

Glaser, B. G., & Strauss, A. L. (2006 [1967]). *The discovery of grounded theory: Strategies for qualitative research*. Aldine Publishing.

Guthman, J. (2008). Neoliberalism and the making of food politics in California. *Geoforum*, *39*, 1171e1183.

Harris, E. (2009). Neoliberal subjectivities or a politics of the possible? Reading for difference in alternative food networks. *Area*, *41*(1), 55–63.

Jaffe, J., & Gertler, M. (2006). Victual vicissitudes: Consumer deskilling and the (gendered) transformation of food systems. *Agriculture and Human Values*, *23*, 143–162.

Matacena, R., & Corvo, P. (2020). Practices of food sovereignty in Italy and England: Short food supply chains and the promise of de-commodification. *Sociologia Ruralis*, 60(2), 414–437.

Mooney, P. (1988). *My own boss? Class, rationality, and the family farm*. Westview Press.

Lamine, C. (2008). *Les intermittents du bio: Pour une sociologie pragmatique des choix alimentaires émergents*. Editions de la Maison des Sciences de l'Homme.

Le Velly, R. (2017). *Sociologie des systèmes alimentaires alternatifs: Une promesse de différence* (p. 200). Presses des Mines.

Lucas, V., & Gasselin, P. (2018). Gagner en autonomie grâce à la Cuma: Expériences d'éleveurs laitiers français à l'ère de la dérégulation et de l'agroécologie. *Économie rurale*, 364, 73–89.

Paranthoën, J.-B. (2013). Processus de distinction d'une petite-bourgeoisie rurale: Le cas d'une "association pour le maintien de l'agriculture paysanne" (AMAP). *Agone*, *51*(2), 117–130.

Porcher, N. (2010). *L'agriculture contractuelle de proximité en Suisse romande* [Mont- pellier. Thèse de Master of Science, Centre International de Hautes Etudes Agronomiques Méditerranéennes, Non Publié.

Renting, H., Schermer, M., & Rossi, A. (2012). Building food democracy: Exploring civic food networks and newly emerging forms of food citizenship. *The International Journal of Sociology of Agriculture and Food*, *19*(3), 289–307.

Rouvière, C. (2015). *Retourner à la terre: L'utopie néo-rurale en Ardèche depuis les années 1960*. PU Rennes.

Shi, Y., Cheng, C., Lei, P., Wen, T., & Merrifield, C. (2011). Safe food, green food, good food: Chinese community supported agriculture and the rising middle class. *International Journal of Agricultural Sustainability*, 9(4), 551–558.

Stock, P. V., Carolan, M., & Rosin, C. (2015). *Food utopias: Reimagining citizenship, ethics and community*. Routledge.

Stock, P. V., & Forney, J. (2014). Farmer autonomy and the farming self. *Journal of Rural Studies*, *36*, 160–171.

Stock, P. V., Forney, J., Emery, S. B., & Wittman, H. (2014). Neoliberal natures on the farm: Farmer autonomy and cooperation in comparative perspective. *Journal of Rural Studies*, *36*, 411–422.

Trauger, A. (Ed.). (2015). *Food sovereignty in international context: Discourse, politics and practice of place* (p. 248). Routledge Studies in Food, Society and the Environment. Routledge.

Tregear, A. (2011). Progressing knowledge in alternative and local food networks: Critical reflections and a research agenda. *Journal of Rural Studies*, *27*(4), 419–430.

Van Der Ploeg, J. D. (2008). *The new peasantries: Struggles for autonomy and sustainability in an era of Empire and globalisation*. Earthscan.

Volz, P., Weckenbrock, P., Nicolas, C., Jocelyn, P., & Dezsény, Z. (2016). *Overview of community supported agriculture in Europe*. Urgenci. Retrieved August 8, 2019, from http://urgenci.net/wp-content/uploads/2016/05/Overview-of-Community-Supported-Agriculture-in-Europe-F.pdf

Vuilleumier, J. (2017). The involvement of community supported agriculture networks in a Swiss popular initiative for food sovereignty. In *Public policies for food sovereignty* (pp. 103–118). Routledge.

Wilson, A. D. (2013). Beyond alternative: Exploring the potential for autonomous food spaces. *Antipode*, *45*, 719–737.

7 Sustainability conventions in a local organic consumer cooperative in Norway

Hope and trouble of participants

Hanne Torjusen and Gunnar Vittersø

Introduction

A more just and sustainable food system is often the rationale and motivation behind the establishment of community food initiatives (CFIs). The local organic food cooperative Vestfold Kooperativ (VK), which will be discussed in this chapter, belongs to a type of initiative that often consists of ideologically motivated people, in this case farmers and consumers who want to collaborate to increase the local supply of organic food. The "alternative", idealistic characteristics and motivations and what unites the actors, which in our context represent the "hope" within the cooperative, are commonly highlighted features of CFIs, while all the trade-offs and dilemmas, which in this context may be referred to collectively as "trouble", are less emphasised (Hinrichs, 2000; Mundler & Laughrea, 2016; Vittersø et al., 2019a). In this chapter we will show how both aspects – hope and trouble – are central to understand the development of CFIs.

It is often assumed that CFIs work as niches that focus on values, norms and conventions, which are different from those that are central to the workings of the mainstream food system. However, recent global crisis with the war in Ukraine, the COVID-19 pandemic and climate crisis have changed the public discourse on food security and provisioning. Sustainability has also come to the forefront in the political debate not least in the EU. Thus, there are reasons to believe that CFIs, like Vestfold Kooperativ, with their holistic perspectives including environmental as well as social and economic aspects of sustainability, may strengthen their role as forerunners and "light houses" for future food system developments. But, as we will discuss in this chapter, CFIs also encounter troubles partly caused by the fact that they are strongly interwoven in the mainstream food system. Their members usually do not depend exclusively on the CFI for their food provisioning. Their relation to the CFI may therefore be described as *hybrid*, as they most commonly operate in several, both conventional and alternative, supply- and provisioning systems simultaneously. This is confirmed by the Strength2Food research project,[1] of which this study has been a part. Like Sonnino and Marsden (2006), we observed a hybridity in the participation of both farmers and consumers in CFIs. They move in and out of different networks, with their own logics and dynamics depending on how the distribution is organised. (Malak-Rawlikowska et al., 2019; Vittersø et al.,

DOI: 10.4324/9781003195085-9

2019a). Farmers can deliver as much of their produce via conventional "long" value chains as via alternative "short" chains (Malak-Rawlikowska et al., 2019). They may be "forced" to use intermediaries and "conventional" actors, either to be able to process their own raw materials or to be able to distribute their products. In the same way, consumers will, typically, procure most of their food in ordinary grocery stores while providing food through alternative distribution channels as a supplement (Vittersø et al., 2019b).

This means that CFIs, as mentioned, do not operate in isolation but rather are woven into the conventional food system which they are opposing. One objective of this chapter is to discuss how this hybridity may create "trouble" as well as correction, inspiration, and "hope" for the development of the CFI and the food system in general. This first line of inquiry in this chapter is related to *outcome*, describing how members perceive the cooperative's contributions to food system sustainability. The second line of inquiry is related to *process*, by investigating the functioning and development of Vestfold Kooperativ.

By this focus, the chapter is attuned to the debate about responsibility in the food system (Wahlen, 2011) and the role CFIs may play in food systems transition (Dedeurwaerdere et al., 2017). CFIs are driven by citizens (farmers and consumers) who actively take responsibility in the food system. At the same time, these actors can most often be considered as the weak players compared to other players in the food value chain (e.g. retailers) and these initiatives are often considered to remain niches in the food system. On the other hand, such food initiatives may play a crucial role in food system transformation by putting pressure on mainstream players of the food system and second, by the role these niche innovations have in broadening the debate about possible and sustainable transition pathways in the food system (Dedeurwaerdere et al., 2017; Laamanen et al., 2022; Carmona et al., 2021).

These reflections apply to Vestfold Kooperativ. It was established in 2015 in the medium-sized city of Tønsberg in Vestfold county, inspired by Oslo Kooperativ, the first Norwegian cooperative of this type. The cooperative only existed for a few years, from 2015 to 2019, and we do not know why it closed down or what impacts the closure had on the cooperative's members and farmers. We know that the farmers continued to sell food through other local distribution channels, such as Community Supported Agriculture (CSA), and continued their local engagement through activities such as on-farm education in collaboration with local schools, receiving the public for farm visits, and organising farm shops, cafés and other hospitality activities. We can only assume that the former consumer-members found new ways to obtain local and organic food.

Community food initiatives in Norway

Community food initiatives are seen as part of a counter trend to the globalised, homogenising food culture (Holm & Gronow, 2019). In Norway this trend is especially prominent in the recent growth of CSAs, market gardens and REKO-rings. Establishment of consumer/producer cooperatives must also be seen as part of this trend. Since the first Norwegian CSA was piloted in 2003–2005 (Bjune & Torjusen, 2005) and formally established in 2006 on the outskirts of the capital, Oslo, nearly 90 CSAs have been initiated throughout the country. The development of CSAs has also spurred the interest in small-scale growing and we have witnessed a significant growth in market gardens since 2019 (Milford et al., 2021). A preferred sales channel for these small-scale producers is a new distribution concept, REKO-rings, that are based on Facebook as a sales platform. This has for many producers and consumers proven to be an attractive way of food provisioning and from its start in late 2017 more than 200 REKO-rings have been established throughout the country (Milford et al., 2021).

In spite of the growth of these CFIs, direct sales is a rather marginal phenomenon in Norway, especially when compared to other European countries (Vittersø et al., 2019b). There are many explanations as to why these initiatives only cover a marginal share of the market. One important factor is the emerging concentration in the food market, with only three major retail chains, that capture 96% of the total sales (Alfsnes et al., 2019). Over several decades specialty stores (butchers, bakeries, fish shops, etc.) have slowly disappeared along with open air markets and similar direct sale. Norwegian consumers have become accustomed to shopping in supermarkets that are numerous, located relatively close to where people live and have convenient opening hours, and where competition for customers first and foremost is about price. Many consumers probably do not consider other sales channels as relevant options for them.

However, the values connected to local, organic food production may fit well with other values discussed as central to Norwegian's identity and way of life. For instance, many informal ways of supplying food, outside the market context, such as growing own produce in the garden, allotments or harvesting from nature, are gaining popularity (Vittersø & Torjusen, 2021). The first "Kooperativ" in Norway started in 2013 in Oslo. Oslo Kooperativ was founded by a group of consumers who wanted to create an alternative sales channel for local, organic and biodynamic vegetables. It evolved into a member-based organisation, where members pay an annual fee and are expected to volunteer throughout the year in order to keep the organisation running. As Oslo Kooperativ has grown, enthusiasm for this initiative has become apparent, as at one time there were over 1,000 people on the waiting list to become members. Several other cooperatives were started in other parts of Norway based on the Oslo model. Although these organisations are located in areas with much smaller populations than Oslo, they operate on the same principles and work to support local organic farmers in their respective regions.

Vestfold Kooperativ started in 2015 on the initiative of a previous member of Oslo Kooperativ's steering group. When she moved to Vestfold county, south of Oslo, and settled on a farm with her family, she found it difficult to access local organic vegetables, despite Vestfold's reputation for being the county that grows the most vegetables in Norway. Building on her experiences at Oslo Kooperativ, she founded Vestfold Kooperativ to enable consumers to access local organic vegetables and support local organic production at the same time.

Vestfold Kooperativ was organised as a non-profit cooperative run and owned by the members. They cooperated with local farmers for direct sales of local organic and biodynamic foods. The cooperative would buy produce directly from the farmers on behalf of its members. By cutting out intermediaries, the cooperative aimed to provide a fair price for the farmers as well as lower costs for its members. Another aim was to lower the environmental impact by less transport, packaging, and food waste. The organisation had a low-budget structure and did not aim to make a profit from their activities; rather, they strived to cut the costs typically earned by agri-food intermediaries as much as possible so that organic and biodynamic farmers earn more from selling their products.

Vestfold Kooperativ consisted of the annual meeting, members' meetings, the board, and working groups. The working groups were responsible for procuring each of the main food products provided (vegetables, meat, and dairy), organising events, logistics, and communication. Food was sourced largely from small-scale local farms, some of which have processing facilities such as flour mills and bakeries, and facilities for dairy and cheese production as well as meat processing. One of the cooperating farms is a large-scale operation that delivered only a small proportion of its production (vegetables) to Vestfold Kooperativ. Deliveries were made via two pick-up points: one in a Salvation Army store in the centre of Tønsberg and the other at the Steiner school[2] in the vicinity of Tønsberg city centre.

The cooperative had a Facebook group with day-to-day communication with its members. Cooperative members pre-ordered bags of vegetables, meat or dairy products from Vestfold Kooperativ's online portal, and paid online or via Vipps, a mobile payment application. Then, volunteers in the coordinating group communicated with farmers about what products they had available and cooperated on creating bags of seasonal products for members. Farmers delivered the ordered products to the drop-off points, where member-volunteers took responsibility for weighing and distributing the products into individual members' bags. The cooperative members would arrive shortly after to pick up their orders. Members committed to either help pack the bags or join a working group. In January 2017, Vestfold Kooperativ had 150 members, of which about 30 ordered frequently (every 14 days). The membership base was strongly supported by the local Steiner school in Vestfold, and many members of the steering group were teachers or parents affiliated with the school.

Theoretical approach: convention theory

In the analysis of hopes and troubles within the cooperative we will draw on convention theory (CT) that originally was developed by the French sociologists Boltanski and Thévenot and later has been modified and applied to food systems research. Boltanski and Thévenot wanted to show how "the human capacity for *criticism* (our highlighting) becomes visible in the daily occurrence of disputes over criteria for justification," and their aim was to "outline a general framework for the analysis of the disputing process in a complex society" (Boltanski & Thévenot, 1999, p. 359–360). We consider CT a useful tool to analyse developments of CFIs as they are, to a large extent, based on criticism or opposition to the conventional food system. It may particularly be related to describing and understanding why certain issues emerge as hope and trouble in our study case, and how the trouble may be overcome and possibly be directed into positive food system developments.

CT has been applied to a number of different topics in agri-food studies in the form of economic analyses of quality conventions ('the quality turn') and in the development of alternative food networks (AFNs) (Murdoch & Miele, 2004; Ponte, 2016; Evans & Mylan, 2019; Andreola et al., 2021). A common assumption is that shared conventions between the actors constitute the strength of CFIs and contributes to their development (Andreola et al., 2021). However, the analytical framework that CT represents is valuable for nuancing and identifying not only conventions that are shared and common but also practices that are affected by different conventions depending on the situation, the outside context, and the role of the various actors within the network or initiative.

Boltanski and Thévenot (1999) developed a grammar of seven common "worlds of worth" (originally six, here including also a green world) organised around different principles or modes of justification (conventions). In the following, we will point to different ways CT might be applied to food (see Ponte, 2016; Table 7.1):

Table 7.1 Overview of "worlds of worth" according to convention theory applied in food-related studies

Worlds of worth and related conventions	
Inspired	Creativity, aesthetic pleasures of food, emotions
Domestic	Trust, family, tradition, relationships, small scale, care provision
Fame	Reputation, recognition, fame, brand
Civic	Fair price, collective, solidarity, justice, public health
Market	Price premium, value, profit, product differentiation, quality
Industrial	Product standard, efficiency, technology, professionalism
Green	Less waste, organic, value of nature, animal welfare

Table based on: Ponte (2016): *Convention Theory in the Anglophone Literature: Past, Present and Future*, and Boltanski and Thévenot (1999): *The Sociology of Critical Capacity*.

Following Ponte (2016) the *inspired world* is characterised by conventions connected to creativity and innovation. In food initiatives, participants' driving, or motivating forces may for instance be their aesthetic pleasures of food. The *domestic world* is centred around familiarity and central values are trust, loyalty, care provision and tradition. It is related to the family, however, in food studies it may also be associated with the social embeddedness of both traditional markets and new food initiatives within local communities. In the *world of fame* visibility and reputation are central driving forces, thus branding and public relations (PR) are important activities for economic actors. The *civic world* is associated with virtues such as solidarity, justice and collective representation and a typical example may be community food initiatives aiming for a more fair and just food system. The *market world* is, following Ponte (2016), centred around competition and profit. This may be connected to price; however, food studies have revealed that especially for small scale and innovative producers, strategies for product differentiation involving a broader set of quality conventions have been successful. The *industrial world* evokes principles of efficiency and standardisation with a focus on technological solutions and with productivity as a measurable goal. It may be argued that the industrialisation of both primary production and food processing with the aim of efficiency and productivity, has happened at the expense of quality and diversity in the food market. The *green world* is the seventh and last "world of worth", that in the wake of the environmental crisis has become more prominent. This is central in alternative food initiatives, but it is also discussed within food studies how conventional players in the food system adopt green conventions in their economic strategies (greenwashing). This exemplifies how the convention theory fits with the critical reparative approach where on the one hand it may be used to identify shared conventions between players in the food system, while on the other hand it may reveal that players may use the same justifications but with quite different and at times opposing goals and intentions in mind.

Data collection

Qualitative interviews

For this chapter, in total nine interviews with ten different persons were carried out from April to October 2017. The interviews included farmers, cooperative organisers and regular members of Vestfold Kooperativ. The farmers were the main distributors of meat, vegetables, and dairy products to the cooperative. Two of the main organisers were also interviewed. One of them was one of the farmers delivering produce to the cooperative, while the other was a regular consumer. We also interviewed three other regular members of the cooperative. Gender representation was even, with five males and five females interviewed. The interviews focused on the motivations, perceptions, and practices of the farmers and members of the cooperative, as well as their perceptions of drivers and barriers for the development of the initiative.

Member survey

The member survey was carried out on two days in August and September 2017 at the two pick-up points where members of the cooperative came to pick up their orders. A total of 28 members responded to the survey. The number of completed surveys was limited by the number of members who had ordered products on these particular dates and because some of the members declined to participate due to lack of time, etc. The intention of the survey was to map some socio-demographic and socio-economic characteristics of the members as well as get an overview of their motivations for participating in the cooperative. The questionnaire was designed to capture reasons for buying from the cooperative that correspond with the modes of justification in the CT framework. Together with the in-depth interviews, it gave a rich picture of the members' motivations and participation in Vestfold Kooperativ.

The survey indicated that members tended to have higher levels of education (tertiary level education) and above-average income. These characteristics indicate that Vestfold Kooperativ may not have been equally accessible to everyone in the local community, which has been raised as a common concern about CFIs. There were more women than men among the survey respondents, not unlike the general pattern for food shopping, and most of the families had children under 18 years of age (Vittersø et al., 2018).

Outcome: holistic perceptions of the cooperative's contributions to sustainability

The most important reasons for acquiring food through membership in Vestfold Kooperativ were support for local producers (civic/domestic conventions), environmentally friendly food production (green convention), and trust (domestic convention) in the enterprise (see Figure 7.1). Access to "high-quality products" and finding the cooperative "innovative and creative" were also among the shared reasons to join the cooperative. These could both be interpreted as belonging to the "inspired" world, while "quality" could also be seen as belonging to the "market" world. Further arguments and judgements related to the market and efficiency (price and convenience) were less important reasons for buying from the cooperative. Participation in the cooperative may for some members have been less convenient because it took up more time, not only for making purchases but also for performing volunteer work within the cooperative. The bag may also have been seen expensive for some members and it was also difficult to compare the price or value of the content of a vegetable bag with a similar food basket from an ordinary grocery store.

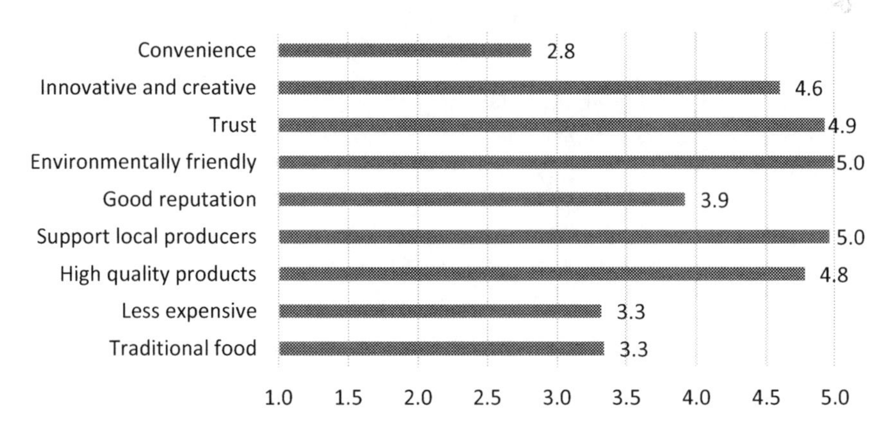

Figure 7.1 Reasons for membership in Vestfold Kooperativ. Average score on a scale from 1 (not important) to 5 (very important). N=28

Previous research on the transformative potential of alternative food networks has focused on the importance of domestic and civic conventions as hopeful for development of these networks. However, some studies have also emphasised that the CFIs are influenced by a plurality of different conventions (Kirwan, 2006; Ponte, 2016). These observations comply with the results of the present study on Vestfold Kooperativ. It seems that these initiatives are motivated by broad issues and seek out holistic perspectives on the food system related to sustainability. In the following we will discuss how this was manifested in hopes and troubles within the cooperative.

Environmental sustainability

The primary aim of Vestfold Kooperativ was to facilitate the consumption, distribution, and local production of organic and sustainable food. In the initial phase, the founder of the cooperative mapped the local area in search of fellow producers of organic food, and a local existing network of interested consumer-citizens was found in the Steiner school community. Vestfold Kooperativ was in that respect founded on a unifying goal of increased organic food consumption.

> All the produce we procure is cultivated in accordance with organic and biodynamic principles. This means any type of farming method that does not exhaust the soil or introduce more synthetic chemicals, fertilisers or pesticide sprays into the earth and the environment. The agricultural methods we support are founded on the desire for healthy and sustainable agriculture, clean water, biodiversity and animal welfare.
>
> (Vestfold Kooperativ)

Several of the consumers emphasised access to locally produced organic food of high quality, for example, with regard to freshness, as an important motivation for becoming a member:

> I prefer organic vegetables and like to get them completely fresh from the producer. . . . and then there is the selection – there isn't a very good selection of organic vegetables around here otherwise.
>
> (Consumer 3)

In addition to increasing the availability of organic food in the local area, Vestfold Kooperativ aimed to "lower the environmental impact by less transport, packaging and food waste".

Economic sustainability

Furthermore, integrated into their economic considerations was care for ethical aspects of food production by avoiding "cheap" food to the detriment of nature, animal welfare, or farmer well-being. It was also a pronounced goal to provide organic food to local households at affordable prices while at the same time securing liveable wages for organic farmers. Reducing or avoiding intermediaries in the food chain, as well as the logistical and practical benefits of the collective efforts of the cooperative, were seen as means to achieving this. These values and views on the economic goals position Vestfold Kooperativ within the civic convention and outside of the conventional economical paradigm that operates within the market convention.

Social sustainability: care in the food system

As described earlier, Vestfold Kooperativ had an identity based on the shared goal of increasing local consumption, and hence production, of organic food. Having such a shared domain of interest constitutes one of the elements of a community of practice, as defined by Wenger et al. (2002). Vestfold Kooperativ could be described as a "community of practice" by the way the members engaged in joint activities, helped each other and shared information (the community element), and developed a shared repertoire of resources, for example, ways of addressing troubles, sharing stories and experiences (the practice element). The shared conventions between farmers and consumers gave directions to the collective efforts within the cooperative and represented a hope that what emerges in practice can inspire further engagement.

This strong community element and the capacity of Vestfold Kooperativ to build local social capital were perceived as a key contributions to the social aspects of sustainability. One way in which the social profile was expressed was in the choice of pick-up point: the Salvation Army shop and café in the centre of Tønsberg. In this way, civic conventions were also closely aligned with domestic conventions emphasising community building.

An emerging value among participants in the cooperative could be described as an "ethics of care", which spanned the food system as a whole and included humans as well as animals, plants and soil. Within conventions theory as applied in rural sociology, care is normally placed within the domestic convention, referring to care for family and within the close relations (Ponte, 2016). What we see here however, includes care for the whole ecosystem on which food production is based. What participants express includes "not destroying" the natural world, but also goes beyond this, in desires to "co-create" with nature, and make a significant positive contribution, as eater or farmer.

A farmer described how he drew inspiration from early experiences with biodynamic farming methods, and visiting biodynamic farms:

> I have become excited about what I have experienced, which appeared to me to be deeply sensible. . . . That is, taking responsibility for being a human being on earth, doing our very best with it without destroying what's around us.
> (Farmer 2)

A similar holistic view was expressed by a consumer-member who also highlighted the collaboration between consumers and farmers, in the sense that the consumers support the farmers who are "doing the work":

> It's important that the farmers get the chance to continually grow the best products, those that are best for nature as well as for people.
>
> (Consumer 1)

The collective efforts of members and what they can achieve together, within and beyond their roles as consumers and farmers, were also emphasised by another farmer:

> Everyone wants to do something for the environment, but few are actually able to do so. But I can do something, I have lots of ideas, and if I can give them a feeling that by buying that bag of food, I can complete a project, . . . That's what I'm really doing.
>
> (Farmer 3)

> I set up birdhouses for endangered birds. Right now, I'm very interested in owls, which are endangered, and I've been very interested in salamanders, the big ones, which are critically endangered, so I've been digging large dams and making small streams for them in the forest. That sort of thing.
>
> (Farmer 3)

A third farmer took this point of potential collective efforts even further when criticising the current food system, taking a holistic and preventive perspective on public health (civic convention) and on collaboration between farmers, medical doctors and consumers.

> Farmers, doctors and consumers should work together. . . . There's only one thing that counts: money, money, money. The consumer is really the one who pays for it: first for the food product and then for getting ill.
>
> (Farmer 5)

This farmer also sought more insights and efforts from natural scientists, while expressing awe and respect for the life-giving qualities of soil.

> I feel we should have had more basic research. . . . It's claimed that there's something like a billion microorganisms in one thimbleful of soil. How many of these do we know the name of? And how they work? . . . It's alive, but we treat it – at least in conventional agriculture – as if it were dead soil.
>
> (Farmer 5)

In his criticism, he draws on both green, domestic, and civic conventions, and his views can be understood as a call for a reconceptualisation of soil as living, similar to what is described by Puig de la Bellacasa (2015). She describes the traditionally close relation of soil science with reductionism, and the present ecological call to engage with soil as a living community. She presents the "foodweb" model of soil ecology for new human-soil relations. Our findings from the Vestfold Kooperativ fit well with a notion of care, which includes relations across the food system – or foodweb. Members expressed care for the living soil, farm animals, each other as farmers and consumers, and future generations.

Process: controversies and solutions

Food production costs versus food prices

Although we found a recognition of the principles of fair prices to farmers and that prices of the bags should reflect the real costs of production, this troubled the cooperative. This issue also turned out as an inner negotiation of the individual farmer: What is the right price of my product? and among consumers: What am I willing to pay for the bag of meat or vegetables? These questions must be seen in the light of the hybrid nature of food distribution and provisioning. The farmers naturally compare prices they get within the cooperative with the prices they can get from distributing through other direct or "long/conventional" sales channels, while consumers compare with supermarket prices where they purchase most of their food. Some consumers were troubled with the price as they considered the different bags very expensive, while others considered them less so, and hoped that they were fairer than organic food bought in the conventional market channels. This suggests that participants did not have a uniform understanding of what organic, local food should cost, or it may reflect personal or household financial circumstances. The anthropologist Daniel Miller explains such negotiations over conflicting moral and ethical issues as a tension between being a loving and caring parent or spouse on the one hand and a thrifty householder/consumer on the other. Being a responsible consumer means to keep expenditures low, thus the price of food will always be an important consideration. This moral norm may also conflict with ethical and environmental concerns if it means to pay more for organic or "fair" products (Miller, 1998).

Similar considerations were also found among the farmers. One of the farmers with a large-scale operation of organic vegetable production, argued that his motivation was to produce organic food at affordable prices: *I don't want them not to buy it because of the price (Farmer 5)*. The content of the bags could sometimes also cause troubles. The price was the same for every bag, but the contents could not always be exactly the same. In this respect, one could say that elements of the market conventions, such as standardisation conflicted with having to handle practical tasks such as sharing a whole carcass in a fair way, without causing any waste.

> They [the consumer-members] are very concerned that it should be just the same in all the bags. . . . But if you take a cow and butcher it, there's only so much tenderloin and sirloin. How do you cut it up evenly? And then there's

> a lot of fuss because somebody has noticed that they didn't get exactly the same as the others.
>
> (Farmer 3)

> It's not supposed to be only about that bag of meat and those sausages. . . . It should be about bringing about change: fewer toxins in nature, and enabling endangered species to live.
>
> (Farmer 3)

One of the consumers gave an example of a situation where he was dissatisfied with the content of a meat bag:

Consumer: So, I actually bought the meat bag last time, because they said it would be lamb chops and leg of lamb steaks and mutton. And it cost five hundred and something. . . . But in these bags it was either or: either lamb chops or leg of lamb, not leg of lamb steaks. . . . But then I think it's a bit expensive, that meat bag, really (Consumer 3).

These quotes show how farmers and consumers basically have different hopes and troubles. Not least, as discussed earlier, consumers constantly negotiate between different concerns. Another issue related to the contents of the bags was that it sometimes caused food waste. Reducing waste was a stated goal by the organisers, and it is likely that they succeeded at some levels, for instance at production and distribution level by getting more of what was produced available for consumption (e.g. by means of flexibility of contents according to season and surplus, allowing for a diversity of shapes). Nevertheless, some of the consumer-members described how they sometimes struggled with finding ways of preparing and eating all of it:

Consumer: But then there are these beets and things that we're not used to, that we wonder a bit about how we can use them. So, it has happened . . . I have thrown food away.
Interviewer: But would you say that you actually throw away more food or vegetables compared to what you otherwise do?
Consumer: No. I don't.
Interviewer: So it's not like you feel like you're getting things you can't use?
Consumer: Yes, there was a time when we got so much red onion. That's not a random example [laughs a little]. And then they turned mouldy and were thrown away. But it's not as if I would ever buy ten red onions anyway. (Consumer 3)

The quote tells us that the bags did not fit the usual purchasing and cooking routines of this family that consequently resulted in food waste.

Overcoming trouble

The cooperative played an important role in communication and facilitation of practical solutions. Regarding the issue of reducing waste, as just described earlier as a challenge, one way in which they facilitated solutions was by providing recipes and tips for how to use the diverse types of produce in the bag via Facebook or the homepage. Some of the consumer-members described how their participation in the cooperative had resulted in new cooking practices, using a wider range of produce, including new combinations in familiar dishes, such as using beets with fish cakes, and sometimes having learned to cook with previously unfamiliar vegetables.

> Kale, for example, which we otherwise would not get much of. There was a recipe for kale-chips which has become very popular since I brought that first bag home. It is really just washing the kale and putting it in the oven, and sprinkle oil and salt over it. . . .The kids are very pleased when there is kale in the bag, because then there will be kale-chips.
>
> (Consumer 3)

Another consumer-member expressed similar experiences of having expanded their ways of cooking:

> It has contributed to new dishes. Twice a week, I make a new type of dish that I otherwise would not have made.
>
> (Member survey, 21)

The cooperative played an important role in this communication. By being "hands-on" in dealing with both farmers and consumers, for example, through regular meetings and joint activities, such as packing food bags, they kept themselves informed about issues of hopeful interest and potential trouble.

> We want to show the consumer how much it [food] really should cost.
>
> (Organiser 1)

The organisers also explained how they sought to mediate an understanding among the consumer-members of why there were smaller amounts of vegetables in the bags when they sourced from a small-scale farmer situated in a less favourable (mountain) production area rather than the biggest large-scale farmer in the cooperative, who also produced for the mainstream supermarket channels, and who could provide the cooperative with vegetables at a lower cost. From a consumer perspective it would be rational to buy all vegetables from the large-scale operation with the lowest price (market convention). Normally, this would have outcompeted the smaller producers and created a conflict between efficiency and volumes (industrial convention) favouring the one big player on the one hand, and the emphasis on fair prices and just returns to all the producers (civic convention) on the other.

The solution was to vary the amounts of vegetables in the bags based on who they sourced from.

Several of the farmer-members expressed how creativity, joy and aesthetic pleasure, central notions within the inspired convention, were important to them as a source of finding solutions and overcoming "trouble" of various sorts. The importance of changes in orientation and value attributions taking place in mainstream society was also highlighted in terms of closing the gap between "the utopias" and mainstream society. One couple described how they had kept up their work at their farm for several decades, starting out with being considered weird and strange, and now realising that the world around them had changed so that what they were doing and had always done "suddenly" was considered "trendy" and evoking hopes.

> It has become so attractive with local food and cultural work in farming, and food is much more in focus – organic food. So, we're more in line with current trends. What was once almost like a parody isn't that anymore.
>
> (Farmer 2, female)

In similar ways, changes, crises, and major events in society, with COVID-19 and the war in Ukraine as current examples, could re-emphasise core values in CFIs, and what used to be considered major troubles may be viewed in a different light.

Concluding remarks: shared and contested conventions

Previous research studies have found that the hopes and troubles associated with food initiatives such as Vestfold Kooperativ rely on the extent to which core conventions are shared among the actors of the group or network involved (Andreola et al., 2021). Aspects of trust that belong to the domestic convention are particularly crucial in this regard. An important strength of the initiative certainly was the close relationships that developed not only between consumers and farmers but also among farmers and consumers, respectively. We also found that Vestfold Kooperativ gained importance beyond the internal organisation and affected areas of the farmers' work and the social lives of the members of the cooperative more generally. This wider hope adds to findings from other research studies, emphasising a more self-referential attitude among CFI/AFN actors (Andreola et al., 2021).

An over-arching shared value within the "green" world was environmentally friendly food production. The cooperative as such was organised based on a common goal of increasing local production, availability and consumption of organic food, in order to achieve a more sustainable food provisioning. Environmentally friendly food production also emerged as important in interviews with producers, organisers and consumers, and received top score in the survey among consumer-members as motivation for joining the cooperative. Animal welfare, minimising waste and generally valuing nature – all values within the "green" world, were also shared between the members of the cooperative.

Within the "civic" world, valuing of the collective character of activities and responsibility, as well as solidarity between producer-members and consumer-members were emphasised. Considerations for health were mutual; producers were highly engaged with producing high quality food which would provide for good health among consumers, and consumers cared about the health and well-being of farmers. Regarding justice and fair price, there was a common understanding of the fairness of covering valid costs of proper and careful food production with organic methods and high regard for animal welfare as well as fair social conditions for farmers. Transparency was the key method for conveying the facts of the production costs within the cooperative; and considerations for achieving "right" prices were expressed from both the consumer-side and the producer-side. Values belonging to "green" and "civic" conventions were negotiated against, for example, uniformity and low price within the "industrial" convention.

A lot of time was spent on processing, packing, transporting, distributing and collecting the various food bags. From a market- and industrial convention perspective these processes can be seen as unnecessarily expensive and very time-consuming. These market and industrial conventions contrast with several of the basic principles on which Vestfold Kooperativ was founded, and where the cooperative wanted to play a role in counteracting societal trends towards larger-scale agriculture, transport of food over long distances, and a situation where the farmer and the consumer never meet. By cutting out intermediaries, food does not necessarily become cheaper for the consumer, but it is a way to support environmentally friendly (green convention) and fair (civic convention) food production. The paradox is that it was precisely related to these conventions that we found the greatest troubles in terms of divided opinions between farmers and consumers. The fact that some consumers emphasise price more than the underlying quality conventions (green/civic) does not necessarily mean that they act on the basis of market conventions. It may just as well be about the conflicting considerations that households face in everyday life between ethics (green/civic) on the one hand and moral considerations (thrift, being economical) on the other (Miller, 1998).

Overcoming trouble – seeking for hope

Vestfold Kooperativ is an example of how troubles may be overcome and dealt with in ways that actually create new hope in solutions and development paths. This applies to the members of the cooperative as well as to alternative ways of organising food distribution more generally. The strength of the cooperative organisation lies in its transparency, where disputes over one convention (issue) are more easily resolved based on other shared conventions and on mutual trust between farmers and consumers. The intermediary, either in the form of the manager/operator of the scheme or in the form of technological/social media tools (Facebook groups, websites, mobile applications) also plays a crucial role in creating and facilitating a unique dialogue between farmers and consumers, solving troublesome issues and thus moving the organisation of the cooperative forward. Being a member of a cooperative also entails a more binding commitment and thus encourages members to find solutions and compromises to the benefit of the cooperative. Even though some of the consumer cooperatives in Norway, including Vestfold Kooperativ, have faced serious troubles, and have even closed down, the networks and experiences are taken into new hopeful initiatives.

We will furthermore suggest an expansion of certain values (the ethics of care) from belonging to one convention (domestic) to arching into the domains of the green and civic conventions. This transformation may be seen as a result of the dynamic interactions between the members of Vestfold Kooperativ, sometimes solving conflicting views and sometimes enforcing shared views and values. These observations may be in agreement with understanding the cooperative as functioning as a community of practice.

Our analyses show that there is a dynamic between shared, hopeful as well as conflicting, troublesome conventions. The hybridity, caused by the fact that participants in the cooperative also take part in food relations within the wider food system, where other types of conventions dominate, is a source of complexity and implies that both consumers and farmers must relate to different and often conflicting conventions. Navigating this complexity is an important key for addressing and overcoming the trouble experienced by the members in the cooperative. The cost of food may be one "unsolved issue" that is a more general question related to social inequality in society at large. In one sense, as the cooperative no longer exists, one may argue that the trouble was not overcome.

Another way of looking at it, however, is that experiences gained through the cooperative may have fed into new forms of hopeful food relations to support local production and consumption of organic food and that the transformation into other forms and constellations may be one way in which CFIs may serve as a hopeful, creative force in the local foodscape. Regardless of how we understand the closure of the consumer cooperative, we find that there are important lessons to be learned and inspiration to be drawn from it in the further development of local and sustainable food initiatives.

Members of Vestfold Kooperativ were clearly motivated by the contributions to food systems sustainability that in their view resulted from the practices and values of the cooperative. These contributions can be seen as related to the broadening of perspectives taken into consideration. The market and industrial conventions did not dominate, they were rather challenged, replaced or negotiated by civic, green and aesthetic conventions. New forms of practices, for example, in cooking and provisioning of food, were formed. New perspectives and ways of understanding were expressed, for example, in the view of human, public health as interrelated with soil health. Expanding views and broadening the debate about hope and trouble in sustainable transition pathways in the food system are some ways in which CFI's such as Vestfold Kooperative may contribute – beyond their limited size.

Acknowledgements

This chapter presents data from the Strength2Food project, which received funding from the European Union's Horizon 2020 research and innovation programme under grant agreement No. 678024. The Strength2Food project brought together some 30 partners across the EU and Asia to investigate, inter alia, the sustainability of SFSCs. For more information, please visit www.strength2food.eu/. The authors would like to thank the participants who responded to the quantitative survey or participated in the qualitative interviews in the Norwegian case study presented in this chapter.

Notes

1 www.strength2food.eu/
2 In English, it is also referred to as Waldorf Schools.

References

Alfsnes, F., Schjøll, A., & Dulsrud, A. (2019). *Kartlegging av utviklingen i butikkstruktur, dagligvareutvalg og dagligvarepriser. (Mapping the development in grocery store structure, product range and prices)*. (Report in Norwegian) SIFO-Rapport nr. 5–19. Forbruksforskningsinstituttet SIFO. OsloMet – Storbyuniversitetet.

Andreola, M., Pianegonda, A., Favargiotti, S., & Forno, F. (2021). Urban food strategy in the making: Context, conventions and contestations. *Agriculture, 11*(2), 177.

Bjune, M., & Torjusen, H. (2005). Community supported agriculture (CSA) in Norway-a context for shared responsibility. In D. Tangen & V. W. Thoresen (Eds.), *Taking responsibility. Proceedings of the second international conference of the consumer citizen network, May 26–27, at the University of Economics, Bratislava, Slovakia* (pp. 277–288). Høgskolen i Hedmark, Oppdragsrapport.

Boltanski, L., & Thévenot, L. (1999). The sociology of critical capacity. *European Journal of Social Theory, 2*(3), 359–377.

Carmona, I., Griffith, D. M., & Aguirre, I. (2021). Understanding the factors limiting organic consumption: The effect of marketing channel on produce price, availability, and price fairness. *Organic Agriculture, 11*(1), 89–103.

Dedeurwaerdere, T., De Schutter, O., Hudon, M., Mathijs, E., Annaert, B., Avermaete, T., . . . Fernández-Wulff, P. (2017). The governance features of social enterprise and social network activities of collective food buying groups. *Ecological Economics, 140*, 123–135.

Evans, D. M., & Mylan, J. (2019). Market coordination and the making of conventions: Qualities, consumption and sustainability in the agro-food industry. *Economy and Society*, 48(3), 426–449. www.tandfonline.com/doi/pdf/10.1080/03085147.2019.1620026?needAccess=true

Hinrichs, C. C. (2000). Embeddedness and local food systems: Notes on two types of direct agricultural market. *Journal of Rural Studies, 16*, 295–303.

Holm, L., & Gronow, J. (2019). Introduction: Eating in modern everyday life. In J. Gronow & L. Holm (Eds.), *Everyday eating in Denmark, Finland, Norway and Sweden: A comparative study of meal patterns 1997–2012*. Bloomsbury.

Kirwan, J. (2006). The interpersonal world of direct marketing: Examining conventions of quality at UK farmers' markets. *Journal of Rural Studies, 22*(3), 301–312. http://dx.doi.org/10.1016/j.jrurstud.2005.09.001

Laamanen, M., Forno, F., & Wahlen, S. (2022). Neo-materialist movement organisations and the matter of scale: Scaling through institutions as prefigurative politics? *Journal of Marketing Management*, 1–22.

Malak-Rawlikowska, A., Majewski, E., Wąs, A., Borgen, S. O., Csillag, P., Donati, M., Freeman, R., Hoang, V., Lecoeur, J. L., Mancini, M. C., Nguyen, A., & Wavresky, P. (2019).

Measuring the economic, environmental, and social sustainability of short food supply chains. *Sustainability*, *11*(15), 4004.
Milford, A. B., Prestvik, A., & Kårstad, S. (2021). *Markedshager i Norge*. NIBIO Rapport nr 7/153. Utfordringer og muligheter med småskala grønnsaksproduksjon for direktesalg.
Miller, D. (1998). *A theory of shopping*. Cornell University Press.
Mundler, P., & Laughrea, S. (2016). The contributions of short food supply chains to territorial development: A study of three Quebec territories. *Journal of Rural Studies*, 45, 218–229. https://doi.org/10.1016/j.jrurstud.2016.04.001
Murdoch, J., & Miele, M. (2004). A new aesthetic of food? Relational reflexivity in the "alternative" food movement. *Qualities of Food*, 156–175.
Ponte, S. (2016). Convention theory in the anglophone agro-food literature: Past, present and future. *Journal of Rural Studies*, 44, 12–23.
Puig de la Bellacasa, M. (2015). Making time for soil: Technoscientific futurity and the pace of care. *Social Studies of Science*, 45(5), 691–716.
Sonnino, R., & Marsden, T. (2006). Beyond the divide: Rethinking relationships between alternative and conventional food networks in Europe. *Journal of Economic Geography*, 6(2), 181–199.
Vittersø, G., & Torjusen, H. (2021). Alternative food supply under the corona. In G. Vittersø, M. Hebrok, N. Heidenstrøm, I. G. Klepp, K. Laitala, T. Tangeland, H. Throne-Holst & H. Torjusen (Eds.), *Sustainable corona life – Changes in consumption among Norwegians during the COVID-19 lockdown in 2020*. SIFO-Project Note 4 – 2021, 25–31. Consumption Research Norway (SIFO), OsloMet - Oslo Metropolitan University. https://hdl.handle.net/11250/2786582
Vittersø, G., Torjusen, H., Laitala, K. M., Arfini, F., Biasini, B., Coppola, E., Csillag, P., Donati, M., de Labarre, M. D., Gentili, R., Gorton, M., & Veneziani, M. (2018). *Qualitative assessment of motivations, practices and organisational development of short food supply chains*. Deliverable 7.1. Strengthening European Food Chain Sustainability by Quality and Procurement Policy (Strength2Food), European Commission, Horizon 2020 Programme.
Vittersø, G., Torjusen, H., Laitala, K., Tocco, B., Biasini, B., Csillag, P., de Labarre, M. D., Lecoeur, J. L., Maj, A., Majewski, E., Malak-Rawlikowska, A., & Wavresky, P. (2019a). Short food supply chains and their contributions to sustainability: Participants' views and perceptions from 12 European cases. *Sustainability*, *11*(17). https://doi.org/10.3390/su11174800
Vittersø, G., Torjusen, H., Thorjussen, C. B. H., Schjøll, A., & Kjærnes, U. (2019b). *Survey on public opinion in Europe regarding contentious inputs – a report*. Deliverable 2.2. Organic PLUS – Pathways to phase-out contentious inputs from organic agriculture in Europe.
Wahlen, S. (2011). The routinely forgotten routine character of domestic practices. *International Journal of Consumer Studies*, 35(5), 507–513. https://doi.org/10.1111/j.1470-6431.2011.01022.x
Wenger, E., McDermott, R., Snyder, W. M., & Valley, I. S. (2002). *Cultivating communities of practice: A guide to managing knowledge-seven principles for cultivating communities of practice*. Harvard Business School Press.

8 The moral economy of community supported agriculture – hopes and troubles of farmers as community makers

Felix Schilling, Stefan Wahlen, and Stéphanie Eileen Domptail

Introduction

Community Supported Agriculture (CSA) schemes have been mushrooming over the past decades. These community food initiatives connect farmers with groups of consumers to provide food whilst building a community. CSAs attempt to reduce farmers' dependency on the global market and to re-establish communal governance over the production and distribution of food, determined by social institutions of the community outside the conventional retail market. But why do farmers need this? Our chapter investigates CSA from a moral economy perspective, considering values and social norms that are attached to turning farming into a community institution (and not only to the outcome of food production; Booth, 1993, 1994), underscoring associated hopes and troubles. *The aim of this chapter is to better understand farmers' perspectives on CSAs, which continuously produce and maintain social relations and enact values through the distribution and definition of roles and responsibilities within the community.* The chapter focuses on the farmers' perspective on CSAs. The social aspect of the supply side of CSA has received less attention than the consumer perspective, even though farmers' willingness to participate in alternative food systems is essential to advance more sustainable food systems. Studies focusing on the producer side of CSA often highlight that farmers engage in CSA for economic reasons (e.g. Cooley & Lass, 1998; Brown & Miller, 2008; Moellers & Bîrhală, 2014; Bloemmen et al., 2015; Hvitsand, 2016; Paul, 2018). Yet, Paul (2018) remarks that aside from economic benefits, ecological and social values also explain the participation of farmers in the CSA. What is still seldom is a depiction from the farmers' perspective of the values of community as an alternative to markets.

Our analysis is based on the assumption that farmers participate in CSA not only to increase their income but also to fulfil their own objectives based on social and cultural values that are not met by the conventional retail market. For example, Timmermann and Félix (2015) address the quality of work of agroecological farmers, demonstrating that farmers strive to achieve meaningful work, a value coined as contributive justice. In this sense, social relations and interactions between consumers and producers contribute to a more fulfilling, pleasant work and living environment for farmers (Bloemmen et al., 2015; Hvitsand, 2016). Also, Forney

DOI: 10.4324/9781003195085-10

et al. (2023, this volume) show how autonomy from the prevailing food system is important for contract farmers. The hope of farmers to increase their autonomy from the conventional food system, can be considered an attempt to escape economic growth imperatives and resist exploitative values. From this perspective, we understand CSA schemes as organisational models through which farmers hope to enact values, which they could not enact satisfactorily in the conventional food system. Building and sustaining a community, paired with questioning and re-creating the producer and consumer roles, is a creative and exciting process, which is instantaneously highly complex, labour-intensive and sometimes troublesome for farmers. In this respect, CSA bares hopeful visions of the future, whilst also demonstrating how troublesome building a community can be.

Moral economy and community-supported agriculture

The moral economy concept has a long history and developed along several traditions (Götz, 2015). The moral economy perspective is an antithesis to contemporary mainstream economic thinking. Neo-classical economists do not necessarily consider the morality of economic action. To a vast extent, they exempt cultural norms and values, rights, and responsibilities or social expectations from their analyses. While all institutions are arguably based on given values, the right thing to do in the neoclassic and dominant economic paradigm is to grow, to maximise profit and increase efficiency. Instead, a moral economy approach acknowledges that economic interactions have moral implications and refers to valuations or norms that can be found in all economic activity (Sayer, 2007). For instance, utility might not be conceptualised in monetary terms only, but incorporates social relationships involved in transactions. A moral economy approach emphasises normative assumptions underpinning economic activity that manifest in the form of rights and responsibilities (Laamanen et al., 2018), hence serves as a point of departure to scrutinise hopes and troubles in CSA. Such a perspective affords to debunk the hopes and troubles being enacted in institutions as well as social and cultural values being associated with CSA (Sayer, 2007). CSA thus serves as a prime example for a re-moralisation of the economy inasmuch normative meanings are associated with production and economic transactions. The moral economy perspective allows us to understand how re-moralising the economy drives farmers and consumers to participate in CSA, enacting their values.

We understand the economic roles and responsibilities of CSA as embedded in a nexus of social relationships, which are structured, and determined by the social institutions of the CSA community. Social institutions are a complex set of social norms organised around values and beliefs of the community members, while social norms are the informal rules operationalising the shared values of the community (Bicchieri et al., 2018; Miller, 2019). Farmers enact their environmental, economic, and social values through building or participating in CSA as an organisational model where values are shared with consumer-members. In CSA, this shared culture usually consists of an intertwined combination of altruistic and political values. CSA relies on values such as reciprocity, solidarity, and fairness, which are different from the values characterising conventional systems of food provisioning (Kloppenburg et al., 1996; Moellers & Bîrhală, 2014; Morgan et al., 2018). Guided by these underlying values, participants set out what is acceptable in their CSA community, what is right and wrong, and how things should be done (Bowles & Gintis, 1998). These values and corresponding goals shape the institutions of the CSA, including the distribution of responsibilities and the definition of roles within the CSA. These roles are a bundle of rights and responsibilities towards the fulfilment of one's own tasks, which in turn are interdependent with the roles of others and the tasks associated with them (Sayer, 2007; Palomera & Vetta, 2016; Bicchieri et al., 2018; Miller, 2019). These various activities and their outcomes contribute to the enactment of the shared values of the community and the personal aims of its members. Depending on whether the consequences of these actions correspond to the expectations of the CSA members and their values, this may lead to the reproduction or changes of the set of social norms and values that constitute the social institutions of the CSA (Palomera & Vetta, 2016; Carrier, 2018). Through their innovation and experimentation, farmers hope for satisfactory outcomes, but they also face troubles and unanticipated dilemmas.

A case study approach

We examine three CSA systems in and around a major city in the German federal state of Hesse (Table 8.1). At the time of data collection, these were the only CSAs close to the city, while there was an estimated total of 25 CSA projects in Hesse. The state of Hesse is characterised by high-quality agricultural soils and historically by small-scale farming structures and family farms. Structural change over recent decades has resulted in around 15,000 farming businesses in Hesse, with an ongoing decline. The current average farm size is around 50 ha and the growth threshold

Table 8.1 Key characteristics of the three CSAs investigated

	CSA 1	*CSA 2*	*CSA 3*
Organisational structure	Farmer-led CSA	Farmer-led CSA including one consumer-initiated sub-group	One farmer in cooperation with three consumer-led CSA groups
Founding year	2017	2017	2012
Pricing mechanism	Bidding round[1]	Fixed price scheme	Fixed price scheme, bidding round
Number of shares	90	230	140
Number and location of depots	One at the farm	One at the farm and three in the city	Four in the city
Farm area	1 ha	60 ha	80 ha
Primary business model before CSA	Salad and herb cultivation for resale	Vegetable cultivation for organic food retail	Vegetable cultivation for own farm store

Source: Authors

below which the number of farms decreases and above which the number of farms increases is around 200 ha (Landesbetrieb Landwirtschaft Hessen, 2020; Hessisches Statistisches Landesamt, 2021). All three investigated CSA schemes started with fewer than 50 members in their first season. Gradually, the primary focus of farm activities has shifted towards CSA and the number of shares multiplied within a few years, with one of them resulting in 230 shares. In fact, the CSA activities on the farms evolved quickly from an experiment to a core activity.

CSA 1 was initiated by the farmer herself and consists of one single group of members. The members hold shares in the output of the CSA production that are subject to a one-year contractual relationship with the farmer. Apart from these contractual agreements, the consumer group is not based on any legally institutionalised structure. Rather, the organisation of the CSA is based on loose and informal relationships between the members and the farmer as well as between the members themselves. All members pick up their vegetables from a single depot on the farm. Prior to the CSA, the farmer specialised in the production of salad and of eight specific local herbs, which are main ingredients in a traditional Hessian dish. The farm produced for one main retailer. On about one hectare of leased land 39 types of vegetables are grown seasonally according to Bioland standards. These include lettuce, cabbage, onions and other vegetables as well as pulses. Almost exclusively seed-proof varieties are used in cultivation. Two full-time horticultural workers and several seasonal workers are employed.

CSA 2 was also initiated by the farmer, but is subdivided into multiple depots and associated pick-up groups. The farmer provides and supplies several depots from where members pick-up their shares. The CSA is set up as a direct contractual relationship between the farmer and individual members. Beyond this contract, there are no organisational structures among the members of the depots and there are no formal relations between the different pick-up groups. After the initiation of the CSA, the farmer was approached by a self-organised consumer group, which was then integrated into the existing CSA structure. In contrast to the other three depots, this group takes care of organisational and administrative matters itself, such as the maintenance of the depot or the administrative management. Previously, the farmer produced primarily vegetables for an organic retailer. On 60 hectares, cereals, and root crops such as potatoes (7 ha), carrots (2.5 ha) and other field vegetables are grown according to Demeter guidelines. A total of 46 different crops are grown. Furthermore, the farmer grows livestock on a small scale with some sheep and donkeys. At the time of data collection, the farm employed six seasonal workers from Romania.

CSA 3 was initiated by a consumer group that approached a farmer with a farm shop and started a collaboration with him. Subsequently, two other consumer groups approached the farmer, and he expanded the CSA scheme. In total, the farmer supplies four depots from three different consumer groups, all three of which identify as independent CSA groups. The different groups take care of their own financial, administrative, and organisational matters, while the farmer provides advice. One of the groups is formally organised as an association, while the other two are rather loose affiliations of people who organise themselves in informal structures. The farm consists of a total of about 80 hectares. 47 hectares are cultivated with different types of cereals, 20 hectares are used for orchards and 10 hectares are used as grassland. Three hectares are used for special crop cultivation, which includes vegetables and strawberries. Furthermore, the farm has 225 laying hens, a herd of suckler cows, horses, and sheep. The farm has been Bioland certified since 2017. During the harvest season, up to five seasonal workers from Poland and Romania are employed. Furthermore, up to three part-time workers are employed for operating the farm shop.

The case study adopted a qualitative approach, conducting an unstructured as well as a structured interview with each of the three CSA farmers (Creswell, 2014). The unstructured interviews, in the form of informal conversational interviews, aimed to gain a better understanding and more comprehensive knowledge of the cases and contexts through largely exploratory questions, while at the same time creating a trusting atmosphere. Subsequently, the structured interviews were based on theory-driven open-ended questions (Honer, 1994; Turner, 2010). The questions focused on farmers' behaviour and work in the CSA, their attitudes about the organisational arrangements in their CSA, the perceived social norms in the CSA community and expectation of CSA members as well as the farmers´ beliefs towards the potential impacts of CSA. Interviews were conducted between November 2018 and July 2019. In addition, the first author conducted observations during on-farm workdays, farm tours, and CSA social events in all three CSAs. Photographs and memory protocols were used for data collection (Esiri et al., 2017; Wagner et al., 2012). Furthermore, the case study integrated newspaper and website articles, as well as internal documents of the CSAs such as harvest tables, member guides, flyers, and membership declarations. All data were stored in a case study database (Yin, 1994). The data were processed and analysed using qualitative data analysis software. Deductive flexible pattern matching was chosen as the analytical approach for the analysis (Sinkovics, 2018; Pearse, 2019). Initially, a codebook of categories based on the moral economy approach was created, including analytical categories about the case context, farmers' values, objectives, and satisfaction towards CSA as well as the social norms and formal structures that shape and provide space for new roles and responsibilities. Subsequently, the data from the individual cases were coded using the initial categories of the codebook. Finally, the results were compared in a cross-case analysis to highlight the differences and similarities between the three CSA schemes.

Results: the making of a moral economy

The making of a community: roles and responsibilities in CSA

Our analysis reveals how farmers attribute roles and responsibilities to themselves and members in the CSA. A continuum of responsibilities held by the farmers and the members range from classical roles and distribution of responsibilities to more innovative ones. Table 8.2 shows how the farmers diverge in the extent to which they take up new roles and share their responsibilities with consumers in each domain of work of the CSA schemes. We found that innovations emerged

Table 8.2 The Domains of Work and Continuum of Reciprocal Responsibilities in the three CSAs

	CSA 1	*CSA 2*	*CSA 3*
Production			
Fieldwork and cultivation	voluntary field workdays as part of farm events	voluntary field work solely as part of farm events	obligatory workdays for members
Planning of cultivation and farming activities	consultative involvement and feedback of members	loose expression of member feedback	active decision and participation of members
Distribution			
Depot management	farmer-led	farmer-led with members' support, member-led in one depot group	member-led
Community Building			
Social activities	farmer-led with the support of the core group	farmer-led with the support of the core group, member-led in one group	member-led with and without the farmer
Communication	farmer-led with support of the core group	farmer-led with support of the core group	core group-led with support of farmer
Administration of CSA			
Pricing mechanism of a CSA share	joint decision for bidding round	Farmer's decision for fixed price scheme	decision of CSA groups, ranging from bidding rounds to fixed price schemes
Finances, membership management, paperwork	farmer-led with support from core group	farmer-led with support from core group	core group-led with support from farmer

Source: Authors

when farmers were willing to delegate some of their decision-making power, and when they took up new roles such as caring for the community or teaching consumers about agriculture. In this context, one strategy is to encompass all members and ensure the highest possible level of inclusion. Such an inclusive strategy also involves those members who link their participation in the CSA only to individual benefits such as improved access to healthy food. This is the case for CSA 1 and 2, which are characterised by low entry barriers. Indeed, these farmers focus on the delivery of fresh seasonal and organic products while requesting little more in return than the upfront payment. In contrast, the selective strategy foresees a greater commitment of members in CSA 3. It involves a shift in the responsibilities traditionally endorsed by farmers and consumers, with the aim to share responsibilities and to reconnect consumers to food production processes. The farmer in CSA scheme 3 supports his members in fulfilling their duties, while the farmers in the other two CSA only started to consider handing over more tasks to the members.

In the domain of *production*, the contribution of members' work to crop cultivation is a key leverage point. The labour contribution ensures that members are strongly committed to a reconnection with food production, which requires members to move out of their comfort zone as passive consumers towards taking over new responsibilities. Thus, the farm workdays point to the desire of farmers to build a community as they reduce the anonymity between farmers and members, create social ties among members and are a locus of teaching and exchange.

> We started with the workdays this year. This involves a lot of work for us, but I also feel that it is important and an essential part of CSA. At the beginning we said we didn't want it because of the workload. But over the last few years I have had the feeling that the distance between the farm and the consumers has become greater, or at least not smaller, and that the tendency towards the vegetable subscription box has become stronger. That was the point when I said: No, I don't want that. Actually, I want that proximity and awareness.
>
> (Farmer, CSA 3)

While regular workdays are mandatory and reflected in the price calculation of CSA 3, they are rather event-like and voluntary in CSA 1 and CSA 2. They were introduced by the farmers primarily to strengthen ties to the farms and promote community bonding. However, the farmers have no intention to make this work contribution compulsory because they do not want to ask for too much commitment from the members. Further, they do not trust members to possess the necessary skills and experiences and thus do not want to be dependent on members' workforce for production, as illustrated by the following quote:

> At the end of the day, we have very good employees who could do everything without the members. So, we are not dependent on them. I don't want us to be dependent on them either because the members from the city are not necessarily capable of helping out here.
>
> (Farmer, CSA 2)

The involvement of members in the production planning also reflects which values are central in the three CSA schemes. The farmer of CSA 3 integrates the members of its three CSA groups into the planning processes. All decisions about cultivation and crop selection are made together. Even new investments on the farm or the introduction of new arrangements such as the working days are discussed and decided collectively. The farmer acts as an experienced advisor assisting the groups in their decisions. To the farmer, granting members this unconventional role and responsibility of co-decision makers on production processes constitutes a learning experience he considers crucial in CSA. He perceives it as the channel to reconnect consumers with the realities of agricultural food production. The other two farmers (CSA 1 and 2) are attentive to suggestions and requests of members, granting them an advisory role. Nevertheless, farmers reserve their right to make the final decision on cultivation and crop selection. The farmers perceive themselves in the role of experts and solely holding the competence for such decisions.

> What is important to me is that the quality of cultivation is maintained. There are also CSAs where people grow their own vegetables and are perhaps supervised by a gardener. You can do that too, but that wouldn't be my thing. I think it's still important to maintain the professionalism of the whole operation.
>
> (Farmer, CSA 1)

The responsibilities in the domain of *distribution* of food are largely determined by the degree of community self-organisation and are not just the result of a decision by farmers. For example, in CSA 2, members who were recruited independently by the farmer or joined on self-initiative may use the same depot, but do not necessarily know one another. In these cases, the sole link among CSA members is their relation to the farm and farmer. The farmer perceives this lacking group awareness as the main barrier to members assuming more responsibilities. As a consequence, the farmer remains quite alone with the responsibility to manage the depots and reports taking over much of the administration of the distribution, although she does not find it desirable. In contrast, responsibilities for depots managed by the self-organised members are distributed differently and these groups tend to take-up more responsibilities in the distribution and maintenance of depots. Consequently, the three self-organised CSA groups of CSA 3 and one sub-group in CSA 2 enable the farmers to concentrate solely on supplying the groups.

> One challenge is certainly to find the depots, but the groups have done that and they also take care of it. And sure, we are part of it, but that's not our job. That's really the groups' job.
>
> (Farmer, CSA 3)

In the domain of *community building*, the role of community maker appeared as a specific new role that the farmers were learning to harness. They are the ones to organise common activities such as farm parties and working days. Yet, here too, the level of self-organisation of the member groups plays a big role in the distribution of responsibilities and the creation of core groups can assist farmers to work towards community. The core groups usually consist of four to five committed members and are existent in all three CSAs with varying degrees of mandates. The core groups are a mechanism for members through which they step into co-producer roles. The core groups take over responsibilities and increase their share in decision-making processes in specific areas such as community activities, communication, finances, and administration. Vice versa, the core groups are an essential structure through which farmers transfer responsibilities and tasks to members, leveraging workload. For example, by engaging in the organisation of social events, members become community makers themselves and support the farmer in transforming local food systems into communities with inner ties.

> Since this year, we have a core group and they are doing quite well. And in the future, I'll try to delegate even more work to it.
>
> (Farmer, CSA 1)

Furthermore, various communication channels and tools like weekly newsletters and websites have been established in all CSAs. These communication and social activities are important to the farmers because they help them to build the community and fall in line with the farmers' desire to reconnect the members to the agricultural realities by promoting and creating mutual acquaintance and understanding. Farmers use communication channels to strive for transparency within the CSA and reciprocate the trust members express towards the farmers through their commitment. In this regard, all three farmers take the leading role but are supported by core groups. Nevertheless, the level of work appears high and farmers report coming to their limits in terms of ability to make time for community repetitively in the interviews:

> We do one or two activities. For example, in spring we planted a herb garden together, where the members can harvest herbs on their own. In summer, the members organised a festival here at the farm and now one of my employees is offering a workshop on fermentation in December. However, I have resolved not to get so involved in it anymore, because it simply exceeds my capacities.
>
> (Farmer, CSA 1)

In the domain of *administration*, two rearrangements in roles and responsibilities took place. First, in CSA 1 and 2, the farmers are primarily responsible for administrative activities such as membership management, the financial processing of membership fees and contractual issues, and receive some support from the core groups. In CSA 3, the three individual CSA groups are largely responsible for the administrative activities themselves but receive support from the farmer if necessary. Second, there are two opposing approaches to the pricing mechanism: the value-driven approach of bidding rounds versus the pragmatic approach of fixed prices. Bidding rounds can be very time-consuming and the higher the number of participants, the more difficult it is to implement the bidding rounds. The fixed price per share approach is faster to implement and easy to administer. CSA 1 opted for a bidding round following a desire for social justice. The farmer of CSA 2 reports that she is reluctant to introduce a bidding round, yet, she expresses a central dilemma. On the one hand, she would like to increase the fairness and inclusion of the CSA scheme. On the other hand, she feels the pressure to get things done and work efficiently with less stress.

> Somehow, I don't feel the need for a bidding round, and I shy away from this act a bit. After all, there are 230 people by now. I actually think it's cool

> because it would somehow raise awareness of the social balance among the people and maybe even more money would come our way because people would be willing to pay more. With 230 people, 230 times either 40 or 70 euros appear on my bank statement. And if I had to calculate with crooked amounts, it would definitely make it much more complicated for me. I mean most of the time you're farming and making sure things are running.
>
> (Farmer, CSA 2)

The making of community includes roles and responsibilities, which might differ along the domains of the CSA. Depending on the setting of the CSA, these roles are coming across in a more traditional or innovative set-up. In any case, hopes and trouble are specific to the domains of responsibility.

Responsibilities and institutional arrangements enacting farmers' values

Table 8.3 lists the values the three farmers pursue and enact thanks to the roles and responsibilities they largely chose and defined in the CSA. Below, we explain what these arrangements are and how they contribute to given values.

The first value concerns the farmers' *agency*. Before engaging with CSA, the three farmers were already pursuing their niche strategies and participating to varying degrees in direct outlets. Although the farmers were able to obtain reasonable prices and financial stability over the years, two of them (CSA 1 and 2) reported an increasing dissatisfaction with the conditions of the competitive market environment in which cultivation and production decisions were largely driven by principles of efficiency, profitability, and quality standards. The key aspects of dissatisfaction were the monotonous work associated with growing few profitable crops, the low value perceived for their efforts and products, and the mental stress triggered by the high pressure to perform. Even though the farmers were able to mitigate the pressure of this system before entering the CSA to some extent, for example, by moving out of the retail business and into direct sales or with better market positioning through specialisation, they perceived that they could not resist or escape this system completely:

> The issue of marketing through trade. Prices are under pressure. The quality must always be perfect. You wonder if that's what you actually want. Do we want more and more, faster and faster? That is also the pressure that is passed on to the employees. Actually, the whole question of growing or abandoning.
>
> (Farmer, CSA 2)

The farmers perceived the collaboration with consumers through CSA as a means to become *less dependent* on the market. The institutions in the CSA, which convey this autonomy and contribute to provide the space for an alternative logic, are foremost the yearly contracts between members and the farmer. These contractual arrangements enable the farmers to transfer parts or all of their ties from the market to their CSA community, as they now sell a greater share of products directly. In

Table 8.3 Values enacted by the farmers and institutional arrangements for realising these values

Values enacted in the investigated CSAs	*Institutional arrangements for realising these values*
Freedom and self-determination	
Autonomy and agency	formal agreement on reciprocal production and consumption provide space for farmers to enact their values
Independence from the market system	contracts ensuring direct sales for the outlet and reducing dependency on market actors
Enjoyment of meaningful work	
Contact with people and connect with consumers	CSA as a direct sales institution with farm-based sales
Reconnect people to food production and farming realities	social activities; planning activities and production decisions, labour contribution through working days
Diversified and multifaceted work and exploring the full range of capabilities	formal agreement on prices which support the production of cost-intensive crops in small quantities and organic production
Fairness and sustainability	
Fairness towards farmers and preservation of local agriculture	upfront payments allow farmers to cover their production costs; local production and consumption based on formal yearly contracts provides secure prices and outlet for farmers
Sustainability through less food wastes and organic food production	consumers accept optically inferior products through tacit agreement about quality of vegetables; reciprocal commitment of diversification and acceptance of available and seasonal crops

Source: Authors

addition, the lesser dependency on outlets in the market appears to create more flexibility and reduce the pressure on farmers:

> If I don't have any lettuce for the reseller, I call in advance and say that I won't have any lettuce next week. And that's just the way it is. Yes, that's how it is with my customers for now. The pressure has been taken out of it quite a bit. If I don't have any herbs, I just don't have any. [These herbs are] only available from me [in the region], and then they have to be happy if they get some.
>
> (Farmer, CSA 1)

At the same time, farmers perceive a strong moral obligation towards the members of the CSA. Farmers admire the good will of members to accept the constraints and related frustrations coming with the delivery of only local and seasonal crops. One farmer expresses the stress of her desire to be fair and reciprocal to the engagement of the members by delivering weekly good products. Thus, the moral obligation can also be perceived as burdensome:

> So, it's more of a burden when I know that next week I'll hardly have anything to deliver. I do search the whole site to see if there's anything left, because I feel that I have to deliver something in return for the money.
>
> (Farmer, CSA 1)

Further, the farmers value the possibility to operate close to their consumers and to *connect* with them. The three farmers work towards increasing the sense of community by actively limiting the size of their groups and organising events enabling contacts among members. This connection enriches their social lives by multiplying amicable relations and even sometimes friendships. Additionally, the farmers aim at reconnecting consumers to the agricultural production by making agricultural realities tangible to the members and to receive appreciation for their work. For the farmer of CSA 3, this even represented the main motivation to start with CSA. All interviewed farmers perceived that reconnecting allows members to become aware of farming activities, and to appreciate farmers' efforts and working conditions. Several rules and agreements in the CSA are designed to meet this aim, including voluntary workdays on the farms, and integration of members into the planning processes of cultivation and the activity of sustainable agricultural production.

Finally, an important aim and value mentioned by all farmers is that of sharing and belonging to a *community*. The idea of reconnecting with members goes hand in hand with the farmers´ perceived desire to build a community that shares the same values and develops a common group identity.

> What I actually liked about [CSA] was that consumers were interested in how food is produced. At this moment I thought to myself that this is actually what I would like to see from consumers.
>
> (Farmer, CSA 3)

> Growing together. Human and nature, but also urban and rural. And yes quite specifically, to know where my vegetables grow, and this is the farm where it is grown for me. Simply this reconnection. I have the impression that this makes people somehow happy. And this is a good thing. Yes, and I think it is quite important to have this connection again. That is almost the most important thing for me.
>
> (Farmer, CSA 2)

In terms of *work quality and the enjoyment of meaningful work*, the outcomes are more two-sided. Although the interviewed farmers generally viewed their activities as fulfilling, they feel a heavy workload. In addition to fieldwork, farm management and other core activities, the farmers are confronted with a variety of new, labour-intensive tasks such as social event management, the organisation of bidding rounds and other time-consuming coordination processes. For this reason, farmers attempt to promote community building so that they can transfer more responsibilities to members. This community making process faces two challenges. First, due to time and resource constraints, the farmers are not always able to catch up with their CSA activities such as community engagement or communications. Second, and more substantially, the interviewed farmers were not necessarily ready to significantly increase their dependence on the community. They feared that this would make them vulnerable while their livelihoods are at stake. One farmer sees working days as a community building measure only and refuses to count on the community for their work. She largely prefers relying on the hired staff, which in addition are skilled. And most importantly, they do not want to lose control over key domains of the farming activity to members of the community or the community itself:

> It was important for me to be in charge of the finances. The idea that someone else would acquire new members and also collect the money. That is, of course, an enormous reduction in workload.
>
> (Farmer, CSA 1)

All farmers reported that diversification of their production as a result of their commitment to deliver a diversity of foods to the community enabled them to diversify their tasks, which affords a more interesting daily work routine. For the farmer in CSA 1, diversification was even the prime objective, in order to escape the monotony of the former specialisation for the retail market. The farmer remarked that she was now growing a variety of crops like cabbage, carrots, and onions that she was not able to cultivate formerly for the retail market because of their low profitability. The demand of the community for more diverse, fresh, and high-quality organic crops pushed the farmers to try out new varieties and innovate, resulting in higher quality work.

Finally, our findings document how farmers and consumers define the new roles and responsibilities to create *fair* working conditions and *sustainable* economic exchanges in the CSA as a local food system. A first element strongly highlighted and appreciated by the interviewed farmers is the contract-based farming with regular payments from the members, which provide a steady income for the farmers and secure wages for their employees. This security reduces the mental stress generated by the uncertainty in the market environment. Furthermore, farmers complement CSA activities with other operations in order to diversify their portfolio. Benefits from synergy effects in their diversified farm portfolios may accrue, such as in the case of the farmer in CSA 3, where all the products are sold via direct marketing channels (farm shop and CSA). In this sense, the farmers see CSA as a suitable means to increase the attractiveness of the profession and to contribute to preserving smaller farms in the region. In addition, the economic profitability and fairness to farmers is strengthened by quality standards being redefined in the CSA to fit the values of farmers and members. Quality in CSA is based upon implicit rules, as norms of acceptance and a strong desire from consumers for environmentally friendly agriculture. These quality standards emphasise taste and sustainable production processes rather than shape, appearance and constant availability. Instead of the institutions of the market, the CSA community takes on the role of setting their own quality standards.

Discussion: transforming roles through reciprocity in the community?

The three farmers of the case studies understand CSA as a means to enact multiple values eliciting hopes and troubles. The community at the centre of the CSAs is grounded on the sharing and enacting of these values. Our results show that the transition from market-based to relationship-based production activities in the CSA leads to a reorganisation of roles and responsibilities. First, through the community, the very roles and responsibilities within these relations between the producers and consumers change to varying degrees in the three CSA schemes. Second, the CSA schemes changed the relations of the farmers to the market and food system.

Achieving community: changing roles for farmers and consumers?

We observed that farmers have a desire for community as a binding mechanism for the function of this collective institution. Also, the community serves as a means to fulfil values related to the farming identity, food culture and cultural landscapes. In both cases, these aims are intertwined through the redefinition of roles of farmers and consumers. For instance, the reconnection of members with the farm and farming activity takes place in all three CSAs through the physical presence of members, picking up the vegetables on the farm, and other means of contribution. In CSA 3, we can even observe how consumers also contribute labour to the production process and become a fixed and calculated labour force through compulsory work days. This points to the hopeful potential of CSA to contribute to the structural change of food systems: the relationship among actors, which might also be troublesome sometimes. Figure 8.1 summarizes graphically the context and type of roles and relationships occurring between farmers and members (eaters) in the investigated CSAs.

When farmers involve members in the farming activities and decisions, they share their responsibilities and thereby give up their role as main planner, as in CSA 3. Here, the farmer claimed to become rather an advisor, counselling the different CSA groups in their decisions on crop choices and amounts, on organisational matters and even on very long-term matters such as on-farm investments. This decision power is accompanied by new responsibilities for members in the management of depots, finances and membership and a work contribution that is necessary for the farm. As the farmer endorses the role of an advisor, the members switch from the role of passive consumers to that of contributors and co-planners. In CSA 3, members co-produce their food in cooperation with the farmer and on his farm. The usual boundary between the roles of producers and consumers in modern agriculture blurs. We interpret that CSA becomes an organisational model and playing

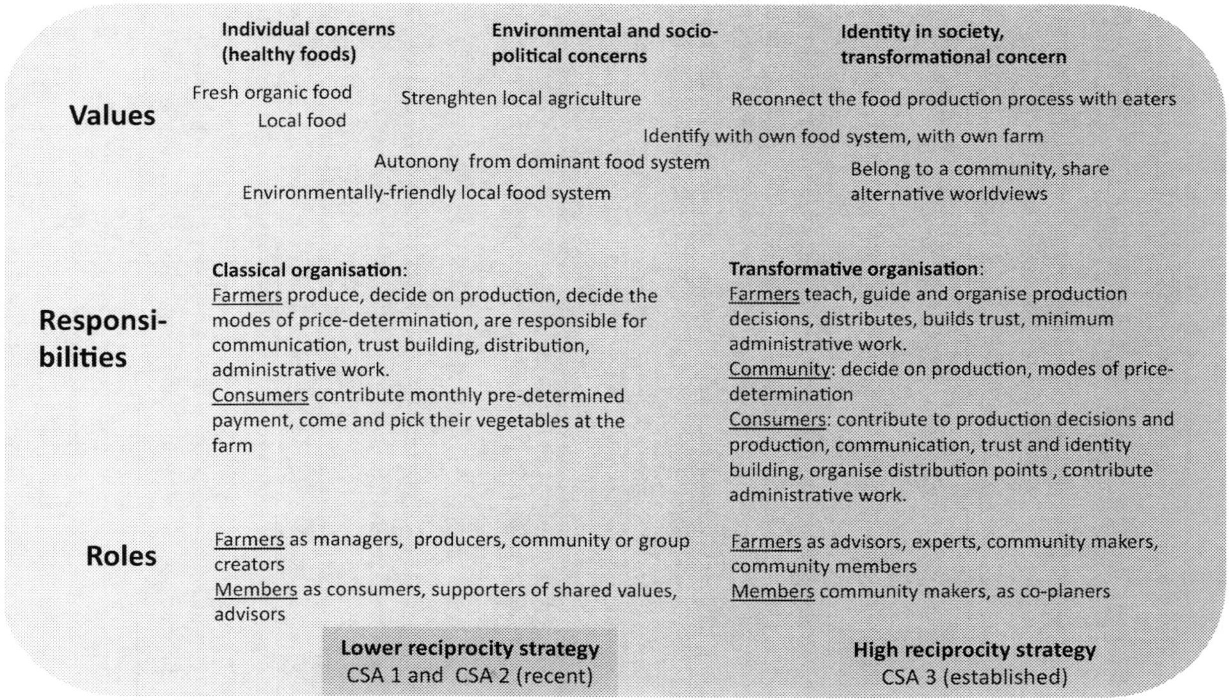

Individual concerns (healthy foods)

Environmental and socio-political concerns

Identity in society, transformational concern

Values

Fresh organic food

Local food

Strenghten local agriculture

Reconnect the food production process with eaters

Identify with own food system, with own farm

Autonony from dominant food system

Belong to a community, share alternative worldviews

Environmentally-friendly local food system

Responsi-bilities

Classical organisation:
Farmers produce, decide on production, decide the modes of price-determination, are responsible for communication, trust building, distribution, administrative work.
Consumers contribute monthly pre-determined payment, come and pick their vegetables at the farm

Transformative organisation:
Farmers teach, guide and organise production decisions, distributes, builds trust, minimum administrative work.
Community: decide on production, modes of price-determination
Consumers: contribute to production decisions and production, communication, trust and identity building, organise distribution points , contribute administrative work.

Roles

Farmers as managers, producers, community or group creators
Members as consumers, supporters of shared values, advisors

Farmers as advisors, experts, community makers, community members
Members community makers, as co-planers

Lower reciprocity strategy
CSA 1 and CSA 2 (recent)

High reciprocity strategy
CSA 3 (established)

Figure 8.1 Values, responsibilities, and roles in CSA – high and low reciprocity in relationships between farmers and members

Source: Authors

ground, in which roles and identities can be redefined, created and tested by the community. It seems important that members see themselves in the role of community makers and contribute to strengthening the community and its underlying values. Sharing the role of community maker between farmers and members seems to be a hopeful step towards empowering members in the food system and an important support for farmers in trying to enrich agricultural production with their own values. Thereby, they contribute greatly as inspiration for new settings in a more just, socially and ecologically sound food system.

However, the aspiration of community building also reveals troubles with CSA. As membership increases, the sense of community and connection suffers. Against this backdrop, the farmers' desire in all three CSA schemes is to strengthen existing relationships in the community rather than expand to new members. At the same time, the rising number of members increases the administrative workload for farmers, which, above a certain size, can no longer be handled by the farmers alone. Thus, for practical reasons alone, a change in the role of members towards community makers, who support the farmer in community building, becomes necessary.

Time may play a role in developing a sense of community and of mutual trust, especially when the groups are large, and the members do not know each other. The longest standing CSA 3 involved members in many more domains and with more depth compared to the other two. CSA 1 and 2 concentrated on the organisation of offer and demand with a more traditional distinction between producers and consumers. It may take some time for farmers to internalise and share new roles and responsibilities with members. The same applies to the members who first have to adopt the values of CSA and learn how the concept works. Ultimately, sharing the same values, such as solidarity, mutual goodwill and friendliness, form the foundation for long-term trust-building, which contributes to a strengthened sense of community. However, building trust and community is a function of the community-building role and requires additional investments of work.

Hopes and Troubles of Reciprocity

The sharing of responsibilities and risks led to a modified relationship between farmers to the consumers, as well as to the market and the food system. Although personal relationships might play a role in the retail market, exchanges are neutralised through monetary payment and not fundamentally inscribed in a context of perduring reciprocal exchange. On the contrary, our results demonstrate how fundamental the perceived obligations and interpersonal dependencies are within the community of CSA. A strong commitment by members to accept seasonal constraints and supply limitations inherent in CSA translates into their formal obligation to buy and eat the farm's limited food supply. If individuals are troubled by these obligations, they often drop out. In turn, farmers feel obliged and hope to deliver the best possible products and to fulfil their part of the contract. Farmers expect members to respect and appreciate their work and efforts. More generally, whether farmers can fulfil their obligations depends on the extent to which members fulfil their own obligations such as paying in time, answering mails, and picking up food. Such interdependencies form the hopes and troubles in CSA because they set expectations for the accepted behaviour of the actors involved. Cooperation, goodwill, and friendliness of the parties involved are basic requirements for mutual exchange in the CSA.

These new dependencies act simultaneously as a relief and a burden for the farmers. On the one hand, the mutual obligations between the CSA farmers and the members lift the pressure of the market by giving farmers agency. The role of the CSA contract with fixed monthly payments and direct sales to engaged consumers reduces risks for farmers and provides a stable income source. As an interesting systematic side effect, the secure income also softens the remaining retail market relationships of the farmers: farmers feel less obligated to the retail market. Moreover, the CSA is a venue for appropriating functions of the market, with farmers and members taking on a role formerly held by the market, which is well illustrated by the quality standards. Farmers and members redefine quality, and appreciate their own standards for quality.

On the other hand, though, the mutual dependency is not perceived as symmetrical by the farmers. Members do depend on the CSA and the farmer for their food, but it is rather an obligation than a dependency, as they have multiple accesses to foods. Farmers, on the contrary, invest their livelihoods in the CSA. Accepting to incorporate the community in the farming process, through more decision-power and work contributions, increases the dependency of the farmer on community, which farmers of the sample were not all ready to do, or even clearly objected to. It appears frightening to depend on members one has little control about. In CSA 1 and 2, the farmers express a desire for a greater community feeling, but it is not a priority. They tend to practise a CSA based on the necessary mutual obligations but with as little mutual dependency as possible. CSA 3 on the contrary is an example where high reciprocal obligations and resulting dependencies exist that are expressed in redefined roles of the actors involved. The farmer accepts to step back as an expert advisor while members act as co-producers. This shows that stronger ties to members go hand in hand with a relinquishment of freedoms and illustrates the trade-off that farmers have to navigate. However, trust and shared values can form the basis for the acceptance of interdependence with which farmers feel comfortable. The willingness of farmers and members to accept and implement their newly defined roles will thus ultimately determine the transformative potential of CSA.

Conclusion

The results of our study of three CSA schemes in Germany emphasise that farmers have been discontent with the conventional food system and are hoping to find alternatives in CSA. We argue that farmers are seeking a means to enact values and create a shared culture around these values. From a moral economy perspective, we investigate how social arrangements and relations in CSA express and materialise specific values and attitudes of the farmers. This perspective uncovers the making of community in CSA, particularly emphasising farmers as community makers. We found that farmers in a CSA structure expand their remit to include non-farming activities and share some of the responsibilities in the distribution and production of food. Thus, new roles may emerge, such as farmers becoming advisers and members co-producers. Importantly, farmers appear also in the new role of community makers in the CSA. The role of farmers as community-makers is a new concept, and the distribution of roles and responsibilities among members is a key factor in designing community food initiatives.

Our analysis reveals the hopes and troubles of social relationships in the CSA: the definition of new roles and responsibilities leads to fulfilling values for the interviewed farmers, and to more meaningful work. These reciprocal relations as well as the organisational model of contract farming, enable farmers to reach economic serenity and reduce agricultural losses and wastes while creating a sense and identity of community. The closed food system around the CSA and the regular contact and involvement of the members in farming directly and indirectly increase the meaningfulness of the farmers' work. The different degrees of mutual obligations between farmers and consumers among the three CSAs can be considered to contain both hopes and troubles. The extent to which farmers are willing to change their role and integrate the community substantially into the planning aspects of the CSA is tempered by their fear of a new dependency on that same community. Farmers are not necessarily ready to depend on a group of farm-outsiders. Yet, trust grows with time and with the feeling and practice of community. Building a community represents a considerable time investment and workload for the farmers, which brings the CSA farmers to their limits. Establishing such as trustful community is not always possible, depending on the places and social context of the farm region. What emerges is a mixed blessing of mutual obligations and dependencies in the CSA communities and a reciprocity that is absent from the conventional market setting. In the end, this chapter corroborates the importance of living laboratories, that is, local initiatives experimenting with more sustainable ways of farming and consuming. In particular, it addresses the costs of building collective values and structures in a predominantly individualistic food system.

Note

1 The concept of bidding rounds is widespread in German CSAs, with the aim of taking into account the different economic circumstances and financial possibilities of the members and thus enabling lower-income households in particular to participate in CSA. For this purpose, the farmer presents the planned cost calculation at the beginning of a new season. An anonymous bidding round takes place, during which members submit their bids based on their willingness to pay for a share. By submitting a bid, each member agrees to participate in the annual budget of the CSA in the amount of the individual bid. The bidding rounds are repeated until the necessary budget is reached.

References

Bicchieri, C., Muldoon, R., & Sontuoso, A. (2018, Winter). *Social norms*. Retrieved November 28, 2020, from https://plato.stanford.edu/archives/win2018/entries/social-norms/

Bloemmen, M., Bobulescu, R., Le, N. T., & Vitari, C. (2015). Microeconomic degrowth: The case of community supported agriculture. *Ecological Economics*, *112*. https://doi.org/10.1016/j.ecolecon.2015.02.013

Booth, W. J. (1993). A note on the idea of the moral economy. *American Political Science Review*, *87*(4). https://doi.org/10.2307/2938826

Booth, W. J. (1994). On the idea of the moral economy. *American Political Science Review*, 88(3). https://doi.org/10.2307/2944801

Bowles, S., & Gintis, H. (1998). The moral economy of communities: Structured populations and the evolution of pro-social norms. *Evolution and Human Behavior*, *19*. https://doi.org/10.1016/S1090-5138(98)00015-4

Brown, C., & Miller, S. (2008). The impacts of local markets: A review of research on farmers markets and community supported agriculture (CSA). *American Journal of Agricultural Economics*, 90(5). https://doi.org/10.1111/j.1467-8276.2008.01220.x

Carrier, J. G. (2018). Moral economy: What's in a name. *Anthropological Theory*, *18*(1). https://doi.org/10.1177/1463499617735259

Cooley, J., & Lass, D. (1998). Consumer benefits from community supported agriculture membership. *Review of Agricultural Economics*, *20*(1). https://doi.org/10.2307/1349547

Creswell, J. (2014). *Research design: Qualitative, quantitative, and mixed methods approaches* (4th ed.). Sage Publications.

Esiri, J., Ajasa, A., Okidu, O., & Edomi, O. (2017). Observation research. A methodological discourse in communication research. *Research on Humanities and Social Sciences*, *7*(20), 84–89.

Forney, J., Vuilleumier, J., & Fresia, M. (2023). Constraint and autonomy in the Swiss "local contract farming" movement. In O. Morrow, E. Veen, & S. Wahlen (Eds.), *Community food initiatives: A critical reparative approach*. Routledge.

Götz, N. (2015). "Moral economy": Its conceptual history and analytical prospects. *Journal of Global Ethics*, *11*(2), 147–162. https://doi.org/10.1080/17449626.2015.1054556

Hessisches Statistisches Landesamt "Auf einen Blick: Kennzahlen zur Landwirtschaft in Hessen" (2021). Retrieved April 16, 2022, from https://statistik.hessen.de/sites/statistik.hessen.de/files/PM_Landwirtschaftliche_Kennzahlen_Grafik_StatistikHessen_28052021.png

Honer, A. (1994). Das explorative interview: zur Rekonstruktion der Relevanzen von Expertinnen und anderen Leuten. *Schweizerische Zeitschrift für Soziologie*, *20*(3), 623–640. https://nbn-resolving.org/urn:nbn:de:0168-ssoar-39274

Hvitsand, C. (2016). Community supported agriculture (CSA) as a transformational act – distinct values and multiple motivations among farmers and consumers. *Agroecology and Sustainable Food Systems*, 40(4). https://doi.org/10.1080/21683565.2015.1136720

Kloppenburg, J., Hendrickson, J., & Stevenson, G. W. (1996). Coming in to the foodshed. *Agriculture and Human Values*, *13*(3). https://doi.org/10.1007/BF01538225

Laamanen, M., Wahlen, S., & Lorek, S. (2018). A moral householding perspective on the sharing economy. *Journal of Cleaner Production*, *202*, 1220–1227. https://doi.org/10.1016/j.jclepro.2018.08.224

Landesbetrieb Landwirtschaft Hessen "Landwirtschaft in Hessen: Ausgewählte Daten und Fakten" (2020). Retrieved April 16, 2022, from https://cdn.llh-hessen.de//unternehmen/agrarstatistik/F_Landwirtschaft_200124.pdf

Miller, S. (2019, Summer). *Social institutions*. Retrieved November 28, 2020, from https://plato.stanford.edu/archives/sum2019/entries/social-institutions

Moellers, J., & Bîrhală, B. (2014). Community supported agriculture: A promising pathway for small family farms in Eastern Europe? A case study from Romania. *Landbauforschung Applied Agricultural and Forestry Research, 3*(64).

Morgan, E., Severs, M., Hanson, K., McGuirt, J., Becot, F., Wang, W., Kolodinsky, J., Sitaker, M., Jilcott Pitts, S., Ammerman, A., & Seguin, R. (2018). Gaining and maintaining a competitive edge: Evidence from CSA members and farmers on local food marketing strategies. *Sustainability, 10*(7). https://doi.org/10.3390/su10072177

Palomera, J., & Vetta, T. (2016). Moral economy: Rethinking a radical concept. *Anthropological Theory, 16*(4). https://doi.org/10.1177/1463499616678097

Paul, M. (2018). Community-supported agriculture in the United States: Social, ecological, and economic benefits to farming. *Journal of Agrarian Change, 19*(1). https://doi.org/10.1111/joac.12280

Pearse, N. (2019). An illustration of a deductive pattern matching procedure in qualitative leadership research. *The Electronic Journal of Business Research Methods, 17*(3). https://doi.org/10.34190/JBRM.17.3.004

Sayer, A. (2007). Moral economy as critique. *New Political Economy, 12*(2). https://doi.org/10.1080/13563460701303008

Sinkovics, N. (2018). Pattern matching in qualitative analysis. In *The Sage handbook of qualitative business and management research methods*. Sage Publications.

Timmermann, C., & Félix, G. (2015). Agroecology as a vehicle for contributive justice. *Agriculture and Human Values, 32*(3). https://doi.org/10.1007/s10460-014-9581-8

Turner, D. W. (2010). Qualitative interview design: A practical guide for novice investigators. *The Qualitative Report, 15*, 754–760. https://doi.org/10.46743/2160-3715/2010.1178

Wagner, C., Kawulich, B., & Garner, M. (2012). *Collecting data through observation*. McGraw Hill.

Yin, R. (1994). *Case study research: Design and methods* (2nd ed.). Sage Publications.

Part 3

Commensality, social gatherings, and food knowledge in CFIs

9 White natures, colonial roots, walking tours, and the everyday

Elaine Swan

Introduction

Somewhat surprisingly little has been written about the colonial, gendered, and raced relations of community food initiatives in the United Kingdom. As the editors of this book underscore, academic writing both celebrates and troubles alternative food networks, urban agriculture, and urban community gardens. In this vein, socialist environmental geographer Salvatore Engel-Di Mauro insists that "the politically mixed bag inhering urban community gardens should give much pause for thought" (2018, p. 1382). Hence in this chapter, I discuss environmental racial, gendered justice through the lens of walking tours of the *Tower Hamlets Food Network*'s seasonal "Gatherings" in London, organised by the *Women's Environmental Network* (WEN) through theoretical ideas on environmental justice, colonial histories, and the everyday.

A 30-year-old not-for-profit feminist environmental organisation, WEN informs women about local and global environmental issues, promotes gendered environmental issues, and supports grassroots food growing movements. Set up in 1988 by women staff from the UK environmental group, Friends of the Earth, to respond to the masculinism of environmental activism, WEN advocates for intersectional gender related environmentalism through practical actions, campaigns, and policy interventions (Buckingham, 2005; Metcalf et al., 2012; Vehviläinen, 2017). For several years, with its roots in ecofeminism, WEN recentred the gendered every day in environmentalism by focusing on the bodily politics of care, reproduction and household labour, and profiles women's interests in environmentalism through their roles as mothers, food providers, carers, and domestic workers in paid and unpaid labour in homes (Buckingham, 2005).

Although WEN facilitated multi-racial food growing activities, until more recently, it ignored broader issues of intersectional environmental justice, that is, an environmentalism that addresses multiple social axes of disadvantage and oppression including race, class, and disability. Second wave ecofeminism, which influenced WEN, has called to task, for over 30 years, activists and scholars of colour for not addressing processes of colonisation, racial capitalism, and systemic racialised gendered oppressions (Mollett, 2017), although some argue this is to misread the multiplicities of ecofeminism (Thompson & MacGregor, 2017). As postcolonial

DOI: 10.4324/9781003195085-12

cultural theorists Ros Gray and Shela Sheikh (2021) argue, white environmentalism excludes the expertise, knowledge, and participation of racially minoritised people in spite of their extra environmental burdens and knowledges. This is also the case in Tower Hamlets (TH) where racialised and other minorities experience the fifth highest air pollution in London and, relatedly, a lack of access to private and green space (Mell & Whitten, 2021)

Indeed, at the same time WEN was established in the United Kingdom, the *Black Environmental Network* (BEN) was founded by environmental justice academic Julian Agyeman, artist Ingrid Pollard, and others. Today, BEN, and other Black environmental groups, advocate against racism in environmentalism, and rural recreation, and for environmental justice and the rights of racially minoritised groups to participate in environmental movements and have equal access to public spaces (Agyeman & Spooner, 1997; BEN).

Since the early 2000s, WEN has encouraged racial, class, and gender inclusivity through its food programme, including initiatives *Spice it Up*, *Taste of a Better Future* and *Gardens for Life*, supporting community gardening groups for British Bangladeshi and Somali women from low socio-economic backgrounds, many of whom are Muslim and victims of racism and Islamophobia. In 2008 WEN established the *Tower Hamlets Food Growing Network*, which today includes city farms, community gardens, first time and experienced gardeners and balcony growers. The network promotes community food growing, biodiversity, sustainability, and recognition of racially minoritised women's cultural, culinary, and growing knowledges as part of their cultural reproduction. With increased attention by organisations to racism in the United Kingdom since the activism of Black Lives Matter, WEN has made issues of race and intersectionality more central to its advocacy, employment practices, and programming.

Whilst acknowledging how WEN draws on intersectional feminist practices of sustainability and food growing consonant with a new wave in environmental politics focused on transforming the everyday (MacGregor et al., 2019), in my analysis of the Gatherings and walks, following feminist and race scholars, I problematise the everyday as an unproblematic category. Feminists argue that the everyday cannot be understood just as background, context, or a given but rather as an analytic category that is historic and invented and unevenly distributed by race, class and gender (Felski, 2000; D. Smith, 1987; A. Smith, 2015, 2016).

Accordingly, the chapter explores how notions of the everyday can be mobilised and troubled by first, reviewing studies on racialisation, nature, and community gardening; second, discussing the walks in urban food and nature environments; and third, surfacing fragments of colonial histories pertinent to the walks to reflect on hopes and troubles. In so doing, I build on current thinking on "everyday environmentalism" in which the everyday is understood as a hopeful "locus for change" to explore how everydayness and everyday environmentalism are unevenly distributed by race, gender, and class (Forno & Wahlen, 2022, p. 434).

To do this, I structure the chapter in a slightly unorthodox way with two mini literature reviews and two mini "findings" sections. As the introduction to the book lays out, a critical reparative approach means fanning hope whilst staying with the trouble, that is, acknowledging things are imperfect and difficult. Therefore, I offer two entrees into the walks, reinforcing this sense of hopeful ambiguity. In holding hope and trouble in tension through the "re-walking" structure of my chapter, I seek to find ways to write the critical reparative in line with the aims of this book. I begin with a brief introduction to TH's racialised, imperial and colonial history as a backdrop to understanding processes, which they produce the everyday, and environmental, activism. These socio-economic power relations continue to structure the present underpinned by colonial and racialised histories distribute everydayness, hope, trouble, and environmentalism unequally by race, class, and gender.

History and context

Towers Hamlets (TH) covers approximately 8 square miles, close to the Thames and Canary Wharf. With a population around 322,704 in 2020, TH is one of London's smallest but most densely populated boroughs. TH's medieval rural landscape origins are in its name, and in those of neighbourhoods such as Stepney Green, Bethnal Green Gardens and Island Gardens (Brown et al., 2019). But brutal colonial histories and legacies, many silenced, underpin many of the borough landscapes and relations for centuries (Ghelani & Palmer, 2021; Wemyss, 2008). The East India Company (EIC) established its docks in TH in 1802, which supported its vast colonial, military, extractivist, and slavery project across the world. Large warehouses, docks, factories, and rows of terrace houses transformed the landscape, and people's lives over the next 150 years (Aziz, 2021; Brown et al., 2019). Through the machinations of the East India Company, TH was seated at the heart of Empire and these histories and ensuing "geometries of power" structure minoritised experiences in the borough and beyond today (Massey, 2005).

By the end of the 19th century, TH was racially and culturally diverse with internal migration from rural England and Wales, major migrations from Ireland, Jewish refugees from Russia and Eastern Europe, and maritime workers from India, China, Malaysia, West Africa, Somalia, and the Yemen who settled in the 19th and 20th (Driver & Gilbert, 1998). The 2011 Census shows 69% of the TH population came from 18 different racial minority groups, 31% white British, and 32% Bangladeshi (LBTH, 2013). Indeed, generations of migrants, refugees, and labourers have defined the changing contexts of the borough, many of whom worked for the EIC or British merchant ships and settled in TH post-World War One working as cooks and peddlers (Aziz, 2021; Rahman & Billah, 2021). But most in these racially minoritised groups confronted by many troubles, including persistent racial discrimination and violence from their arrival to today (Aziz, 2021; Rahman & Billah, 2021).

During the Second World War, bombing devastated many buildings and homes in the Borough. The post-war period saw the decline of the dock industries with entrenched deindustrialisation, rendering substantial areas of land and buildings derelict (Brown et al., 2019). To address labour shortages following the war, migration from former colonies was actively encouraged through changing immigration laws, with many settling in East London from Bangladesh and other Asian and

African countries to find a better life, employment and avoid conflicts (Aziz, 2021; Brown et al., 2019; Rahman & Billah, 2021). Job opportunities mainly included employment in low paid sectors, with unskilled and semi-skilled work (Rahman & Billah, 2021). Moreover, the consequences of imperialism and colonialism, deindustrialisation, under-funding and migration affects many residents of TH, as social theorist Abdul Aziz stresses,

> All these groups have historically been the subject of unjustified abuse, scapegoated, and apportioned blame for the historical and current ills of the host society, despite contributing little to the perceived problems of the host society and often living in abject poverty.
>
> (2021, p. 7)

Furthermore, migration scholar Georgie Wemyss (2008) argues that museums and the media erase the brutal histories related to the EIC imperialism and colonialism in TH. For instance, media and scholarly works neglect the centrality of the enslavement of Africans and profits of slave owners and traders to the existence of the docks. As a result, local African Caribbean working class residents are not seen as descendants of dock workers, who are often constructed as white working-class. Furthermore, the EIC and the British government's roles in the many devasting famines in India and destruction of peasant food ecologies are erased (Damodaran, 2015). Outside of the Bangladeshi community, few know the targeted killing of an estimated three million Bengalis by the Pakistani Army and allies in the 1971 war, connected to the British division of British India into sovereign states of Indian and Pakistan in 1948, eventually leading to the establishing on an independent Bangladesh (Sharma, 2020). To avoid these brutal conflicts and the ensuing political turmoil in Bangladesh, a large number of people immigrated to the United Kingdom (Rahman & Billah, 2021). But in spite of many war atrocities, Bangladeshi people who escaped to the United Kingdom in 1971 are not constructed as refugees "fleeing persecution" (Wemyss, 2008).

These histories have left their material, social and cultural scars including persistent racism and Islamophobia. As historian Jane Jacobs writes, "the foundational ideologies of imperialism live on in this city, shaping contemporary economic status, local class divisions and racial politics, and nationalist articulations" (cited Driver & Gilbert, 1998, p. 22). Consequently, racist groups carried out repetitive racist and Islamophobic attacks on racially minoritised groups, especially on Bangladeshi families and their businesses in TH in the 1970s, and post-9/11. In the 1970s, property was vandalised, windows were smashed, doors were smeared with excrement, and children, men, and women were physically assaulted (Rahman & Billah, 2021). As Mahfuzur Rahman and Morsaline Billah write of the 1970s attacks: "These incidents left the Bengali families intimidated in such a way that they felt insecure to leave their houses . . . Women walked . . . in groups to protect themselves from potential violence" (2021, p. 200). Since 9/11, white supremacists have attacked girls and women wearing hijabs going about their everyday business in TH and other parts of the United Kingdom.

Today, TH has higher rates of poverty, infant mortality, unemployment, and early death than most other London boroughs (GLA Intelligence, 2015). National and local neoliberal capitalism, "rampant financialisation," austerity, intensive "regeneration," state-led gentrification, and COVID-19 continue to deepen inequalities (Le Grand, 2020). Urban neoliberalisation and gentrification created new forms of racial gendered spatial relations with minoritised groups, especially working class white and racially minoritised women, affected by the increased cost of living, social reproduction, limited housing, and reduced democratic influence, breaking up community social networks which traditionally supported women (Curran, 2017; Listerborn, 2020).

While these inequalities do not define people's lives, cultures and aspirations, they do play out economically, socially, and ecologically in everyday lives, including access to green spaces, of relevance to WEN's food growing and environmental initiatives. Environmental planners Ian Mell and Meredith Whitten (2021) show that economically deprived boroughs like TH have fewer accessible, high-quality public green spaces with Black communities four times less likely to have access to outdoor space. Disadvantaged communities receive less public spending on "green infrastructure" and over 70% of TH schools had less than 10 m^2/pupil of green space.

Of course, racially minoritised groups have found ways to survive and resist the multi-layer exclusions, building networks, community groups, social support and gaining a strong foothold in politics, business, teaching, the public services and the law. They have fostered the means build hope and to stay with multiple kinds of trouble. My all-too-brief recounting of these colonial gendered and raced histories underlines how hope and trouble are unevenly distributed.

Everyday environmentalism, race, and gender

The previous section described the racialised and colonial history of the TH landscape, where WEN's food-related everyday environmental activism takes place and shows how "everyday environmentalisms" are therefore structured by histories, racisms, migrations, and inequalities. Environmental scholars Francesca Forno and Stefan Wahlen (2022) refer to the term "everyday environmentalism" to point to the way that the "everyday", variously understood as the familiar, routine, and mundane, has become an important site for environmental change in practices and narratives. Using insights from feminist Rita Felski (2000), they show how the everyday in environmental politics and practices eventuates through facets of the everyday – space, time, and modalities: domestic space of the home, different rhythms through slow food, slow livings, de-growth – and new habits and social imaginaries, respectively.

But the everyday is not quite as self-evident as sometimes thought. As feminist and critical race theorists argue, it is an invented analytic category through which race, class, and gender are constituted (Felski, 2000; D. Smith, 1987; A. Smith, 2015, 2016). Although feminists have attended to the gendering and classing of the everyday, they give much less attention to race. In this way, many theories of everyday life are complicit with how racialisation is made ordinary (Smith, 2016).

In contrast, drawing on the work of 20th-century pioneering Black scholars such as W.E.B. Du Bois, bell hooks and C.L.R. James, Andrew Smith argues that "race weaves into and out of our understandings of the quotidian, the ordinary and the mundane, including how these terms are used by sociologists" (2016, p. 26).

On this view, ascriptions of "everydaydayness" reproduce notions of perceived racialised difference. The everyday in the concept of "everyday racism" underlines the everyday as context for racialisation and racism. But additionally, Smith argues that the mundane, ordinary and habitual are normatively white. Being included or excluded from the "condition of everydayness" and being "part of a merely mundane world" are how racialisation happens (Smith, 2015, p. 1145).

For racially minoritised people, everyday life cannot be merely routine, habitual or characterised by absent-mindedness or living in the moment that theorists such as Henri Lefebvre and Michel de Certeau associate with the everyday. This is because for Black people as James argues, their ordinary practices are "violently regulated" (Smith, 2015, p. 1145). "Whiteness comes to epitomise the everyday quality of taken-for-grantedness, the norm, the mundane, the self-evident" (Smith, 2016, p. 62). Such thinking raises questions about how everydayness, and everyday environmentalism can be unevenly distributed by race, class, and gender.

Racialisation of nature

Some of these understandings are implicit in studies on environmental justice, nature, race, and community food gardening. Scholars and activists have discussed the racialisation and racism of nature, environmentalism, and alternative food networks for some time (See for instance, Agyeman, 1990, 2014; Agyeman & Spooner, 1997; Byrne, 2012; Matless, 2016; Slocum, 2008; Guthman, 2008a, 2008b). More recently, growing scholarship and activism on race, colonialism and environmental justice challenges white centric understandings and practices, especially in what is considered everyday domestic gardening (inter-alia, Czennia, 2021; Fowler, 2020; Ernstson, 2020; Hickcox, 2018; Ratinon & Ayre, 2021). What's significant in this work is the view that environmental justice and environmental resources are racialised in place-specific, historical ways, through institutional structures, and relations of colonial and racialised power (Agyeman et al., 2016). Such work points to the power relations underpinning landscapes, land use, and constructions of rural and urban natures, which inflect how differently racialised groups access green and food growing spaces.

The construction of nature produces inequalities and exclusionary practices on an everyday level (see, for instance, Byrne, 2012; Matless, 2016). Work on rural nature shows how the associations between rural landscapes, English nationalism, white histories and values, position the racially minoritised as Other and code certain natures as national white spaces (Agyeman & Spooner, 1997; Matless, 2016; Panelli et al., 2009). In the United Kingdom, racially minoritised groups experience racism visiting the countryside; their everyday presence in rural villages is erased, and their ecological knowledges ignored. Even though many immigrants to the United Kingdom from South Asia, Africa and the Caribbean have rural and farming backgrounds (Agyeman & Spooner, 1997). In similar vein, Corinne Fowler (2020) points to the erasure of the historic contribution of Africans, Indians, and Chinese to the shaping of garden cultures and botanic gardens in the United Kingdom and the ex-colonies. As environmental feminists, Charis Thompson and Sherilyn MacGregor put it, ideals of nature are "steeped in deeply sexist, racist and transnationally unjust imaginaries", with notions of "pristine nature", used to discriminate against racialised and LGBTQI people (2017, p. 49).

The racialisation of nature extends to urban natures, everyday practices and notions of environmentalism. Writing of migrants' everyday relations with nature in the city in the United Kingdom, environmentalists Clare Rishbeth and Jo Birch argue that "too often nature – rural, wild and urban – has been predominantly articulated by White middle-class voices . . . and sometimes with disturbing colonial or exclusionary underpinnings" (2020, p. 11). Talking of a US context, geographer Jason Byrne (2012) shows how "White" or Anglo-normative ideas/ideals of nature make urban parks inflecting their design and layout, and facilities and leisure programmes. Parks become the spatial and material expression and legitimation of nature-based ideologies and practices and reproduce racist and/or elitist power relations. As a result, he argues, environmental and health benefits garnered through everyday practices such as spending time surrounded by trees, away from the noise and air pollution of busy road networks are unevenly distributed by race, class, able-bodiedness, and migrant status. Research from 2019 in the United Kingdom shows that 25% of racially minoritised people "over 16 never visited a park or other natural site" and "only 56% of children spent time outside once a week" (Walton, 2021, p. 15).

Environmentalism, alternative food, and whiteness

Mainstream environmentalism has been an exclusionary white and middle- or upper-class social institution, reproducing a dominant middle class white environmental subjectivity, or habitus (Gray & Sheikh, 2021; Hickcox, 2018; Finney, 2014). Black geographer Carolyn Finney (2014) argues that processes of racialisation and cultural representation affect who participates and is heard in environmental debates. Environmental scholar Abby Hickcox (2018) argues that racially minoritised group's environmental ethics and practices often lie outside of mainstream white norms. Consequently, white environmentalism excludes the expertise, knowledge, and participation of racially minoritised people, creating environmental racism, narrow ideas of natural resource management and white-centric sustainability practices. In this way, certain spaces and natures are ordered, organised, and accessed by race and class: racially coded and accessible according to environmental privilege (Brahinsky et al., 2014).

In a similar vein, scholars and activists critique the alternative food movement for racism, racialisation and (settler) colonial logics which reinforce racial hierarchies, and entrench marginalisation and dispossession (Alkon & McCullen, 2011; McClintock, 2018; Ramírez, 2015; Slocum, 2008). Community gardens, urban farms, and farmers markets are seen as white spaces which, while claiming to be inclusive, reproduce exclusionary practices (McClintock, 2018; Slocum, 2008). For instance, Alison Alkon and Christie McCullen (2011) argue that a white farm imaginary perpetuates white privilege and constrains participation and anti-racism. Critical food studies scholar, Julie Guthman proposes that "a set of discourses that derive from whitened cultural histories" undergird the US alternative food movement reproducing ideas about minoritised groups as passive victims (2008a, p. 434).

Such scholarship underlines that urban agriculture was not "pioneered" by white people as is often represented, but historically practised by racially minoritised and other low-income groups to supplement their diets (Chou, 2018; Reynolds & Cohen, 2016). This critique calls for recognising the erasure of these histories but also for the valuing of the expertise of minoritised groups and an interrogation of the promotion of white people as creative, entrepreneurial and adventurous and people of colour as backward and in need of help (McClintock, 2018).

While race and colonialism are increasingly being discussed, gendered relations are often missing in these studies. A stark neglect given that gender plays a significant part in environmental injustice and women environmental activists and women led movements have played a crucial role in creating ecological awareness about the food system (Buckingham, 2005; Thompson & MacGregor, 2017). Gendered power relations structure the distribution of access to and control of natural resources and the division of labour (Jarosz, 2011; Martin, 2019).

But gendered racialised politics also undercut environmental activism. For instance, Gender studies scholar, Sally Rifkin (2018) argues that the hypervisibility of white women in food justice projects, means that contributions made by black women are invisibilised. Today, there is an increasing feminisation of community gardening in the United Kingdom. Susan Buckingham (2005) suggests in the United Kingdom racially minoritised women are increasingly growing food on urban allotments and in community gardens. In this vein, feminist scholars underline how food growing can be understood as an act of feminism through the ownership of production, regaining economic control, placing importance on small-scale environmentalist actions, and the valuing of transnational racialised food knowledges and sharing such that WEN supports (Buckingham, 2005; Metcalf et al., 2012; Vehviläinen, 2017).

Walking as environmental activism

In this next section, I draw on my participant observation of guided walking tours at WEN seasonal outreach events in TH, called Gatherings. Since 2010 the Gatherings have run quarterly in Spring, Summer, Autumn and Winter (except for during the COVID-19 lockdowns) the events are designed to be convivial, commensal and pedagogic. The Gatherings reach a wide audience of between 80 and 100 people, mainly women but with a few men and several children, all of different racial, ethnic, faith, and age backgrounds, mental health status and able-bodiedness, and mostly of lower socio-economic status.

Each Gatherings is themed – for instance, biodiversity or wellbeing – and comprise talks, experiential workshops, a shared lunch, the much-loved raffle, and a guided walk in a "green space" in TH. Often led by women of colour, who share their expertise, the experiential workshops covered topics such as making composts, wormeries, and home-made cosmetics. The workshops are designed to provide embodied, visceral, tactile, and material engagements with plants, soils, composts and foods to help us reflect on our everyday food provisioning habits. Using hands-on approaches, they enable attendees to develop and share sensory skills and knowledge and facilitate experiences of the processes of material transformation of ingredients and materials (Hall et al., 2020). Participants undergo bodily encounters with each other, and the group transforms as attendees sit close together, observe each other, chat, laugh, and ask questions. Ecofeminist literature stresses that women's bodies encounter environmental inequalities through domestic and caring work and at the Gatherings they are centre stage, invited to play and experiment (Buckingham, 2005). Through their inclusivity and these processes, the Gatherings enable a women-centred, multi-racial and differently bodied "viscosity" – a collective of bodies sticking together – which differs from many white middle-class food and environmental activities (Ramírez, 2015).

The walks build on and extend the sensory, bodily, and affective registers of the other activities. They are designed so that attendees can visit edible landscapes and vegetal spaces such as parks, community gardens, and urban agricultural sites in the borough. Typically, local experts from the sites lead the guided walks. The walks offer the potential to be in, and move through varied urban green spaces: streetscape, parkland, woodland, graveyard, and urban farmland. The walking

Figure 9.1 Intercultural and generational mixing as attendees walk around the Cemetery Park (Author credit)

Figure 9.2 Walkers encounter the Bangladeshi vegetable kodu grown outside of an estate, a glimpse for some of a different everyday vegetable (Author credit)

facilitates physical, sensate, and kinaesthetic movement of the body through these urban natures and an opportunity to socialise and connect with humans and plants, trees, insects, birds, (O'Neill & Roberts, 2019). Through the tours, people can explore cultural, agricultural and planty complexities and the materiality of the green spaces. People are encouraged to develop their "planty attention" and sensorial practices, sniff, and smell – and even taste on the foraging tour – immersing their bodies sensorially, physically, and emotionally (Pitt, 2015). The walks enable a shared encounter with place, inter-racial and intergenerational mixing, unexpected conversations and encourage embodied and emotional ways of knowing nature, food and the city, outside of mundane, familiar pedestrian practices (Walter, 2013; Robinson & McClelland, 2020; O'Neill & Roberts, 2019).

Walking with goats

Here I discuss in more detail the guided tour around Stepney City Farm during the 2018 Autumn Gathering themed on City Food. A three-acre urban farm, Stepney is sandwiched between the major loud, busy, and polluting roads of Mile End Road and Commercial Road, situated off Stepney High Street and Garden Street. The financial district of Canary Wharf glowers from the East; the Gherkin, a famous office building, housing transnational companies Swiss Re and Sky News, and the City of London in the west; and behind the café, St Dunstan's and All Saints Church. The site was derelict and vacant in the 1970s, having been bombed like much of the East End in the Second world war. Local residents converted it into Stepping Stones urban farm in 1979 just as urban farms were being established (Sankey, 2015). Due to lack of funding, it fell into disrepair until 2010 when it was revived by local volunteers as Stepney City Farm.

An educational charity, Stepney City Farm houses artists, two classrooms, a small Farmers Market every Saturday, local resident allotments, barns, farmyard animals, and a café. Its aims are to improve "lives through farming; we provide a welcoming green space to bring together our diverse community, and cultivate wellbeing through high-welfare, environmentally-sustainable farming practices" (Stepney City Farm website). The farm grows and sells produce including organic meat, eggs, flowers, and vegetables, and taking account of the dietary needs of Muslim neighbours, the meat is Halal. Free to visit, the farm offers a varied education programme including school visits, classes, training activities, and tours, with over 500 children and young people visiting each year.

On the day of the Gathering, as I walk off the nearby roads, the Farm immediately feels like a haven. I notice unruly flower beds, tumbling nasturtium, with goats playing on the left, and a wooden classroom pod, the venue for the Gathering further towards the allotments. In the afternoon, eight of us do the walking tour around the farm. We start and stop as we encounter different vegetal landscapes, animal and plant agricultures, and material objects such as sheds, fences, and compost heaps which form a patchwork of growing and composting spaces, and different natures. Nestled between low and mid-rise council flats, and three busy roads, the corporate towers of Canary Wharf on the near horizon, with the cockerel crowing, the rustle of chickens and pigs eating, and the smell of animals, compost and their excrement, the farm has the potential to create urban human and non-human relations in multiple emotional, sensory, embodied and cognitive ways. The Farm represents animal and plant agriculture in a highly urbanised and impoverished borough.

As we amble around the farm, our guide offers us the opportunity to connect with food, farm practices, farm animals, different types of plants, and lands in embodied ways to develop our ecological and food growing attention. The tour elicits a range of tactile and emotional responses: touch, smell and even taste which augment the sights and sounds of the farm. We meet farm animals: food animals such as pigs, goats, chickens and petting animals like the ferrets. The "food animals" will be slaughtered. Our guide does not go into this in detail but intimates by mentioning the meat sold in the shop. The walk provides the potential to reconnect to "the-situation of abstract food 'products' into relational contexts of place, people, and process" (Johnson, 2018, p. 1664).

From almost every angle of the farm, we can see the allotments. There are 64 allotments designed and tended by local residents of different racial and ethnic backgrounds – white British, and British Bangladeshi and African Caribbean. What's immediately striking is how varied the allotments are: little micro-territories with their own planty designs and shed-cane-netting architectures. Attractive by standard garden aesthetics and untidy for others, the plots represent diverse material and symbolic cultures of gardening, sustainability, and food.

The recycled canes, branches, water-bottles, plastic bags, and other paraphernalia create everyday environmentalisms, crafted through the embodied skills, creativity and daily practices of the gardeners. Diverse in plantings and structures, the plots are mostly for food production and grow different types of "ethnicised" vegetables and fruit with a whole assortment of sensory characteristics. The allotments represent racialised, gendered, and classed produced natures supported by the gardeners' hopes, interests, knowledges, motivations, and everyday practices (Domene & Saurí, 2007).

The growers' digging, weeding, planting and harvesting offer an alternative challenge the whiteness of alternative food and farm imaginaries and community garden bodily viscosities (Alkon & McCullen, 2011; Ramírez, 2015). Environmental and gardening knowledges and practices of people from minoritised groups are often erased in dominant popular media and culture but at the farm, these are materialised, sprouting, blossoming, climbing, seeding, and taking root, and in full view on the walks (Fowler, 2020; MacGregor et al., 2019; Rishbeth & Birch, 2020). At the same time as these practices reference other landscapes, soils, scents and culinary cultures, the allotment users make East London urban landscapes more inclusive. As such their sensory, material and symbolic ecologies tell those of us on the walk a different story from those "discourses of alternative food [which] hail a white subject and thereby code the practices and spaces of alternative food as

Figure 9.3 Allotments (Author credit)

Figure 9.4 Everyday environmentalisms in the allotments (Author credit)

white" (Guthman, 2008b, p. 388). Moreover, they enable sensory encounters with hybrid, cosmopolitan natures in semi-public and public places (Ageyman, 2014; Gandy, 2022). Like the kodu and chillis growing in allotments and gardens in TH, they represent by place and home making by transnational and multi-racial communities from the former colonies. And they offer multiple potential hopes, against entrenched history colonial, and contemporary racist and sexist troubles in that they are

> cultural and community inscriptions on the cityscape that offered a statement of presence, of recognition, that both humans and nature(s) in cities are becoming increasingly different, diverse, and cosmopolitan, and are welcome.
>
> (Ageyman, 2014, p. 21)

These memorialise transnational, intercultural, sensory, environmental and culinary pasts, identities and practices (Tolia-Kelly, 2004). But as Ageyman insists, they enact:

> greater difference, not only in human but also in ecological terms. Viewed in this way, the urban area becomes a socionatural system comprising a wide

> range of life forms, cultures, and possibilities: it becomes an intercultural city ecosystem offering inclusive possibilities.
>
> (Ageyman, 2014, p. 24)

Furthermore, these "cosmopolitan urban ecologies" have possibilities for food provisioning in the future (Gandy, 2022). For instance, Bangladeshi and Caribbean vegetables have the potential to offer environmental benefits and food security in the face of changing United Kingdom ecologies with climate change (Kell et al., 2018).

Depending on the histories and racial backgrounds of those of us on the walk, these culinary and growing cultures will seem exotic or everyday, but whichever, the allotment gardeners on the farm claim the right to take greater control of urban space, becoming "producers of natures" not just consumers of natures (Domene & Saurí, 2007).

Everyday hopes

Through the Network and its outreach, WEN invites those most marginalised and traditionally invisibilised to participate in feminist environmental and food politics. Feminist in the sense that their various activities from food growing, advocacy and Gatherings profile women's bodies, expertise, and practices. The Gatherings and walks represent shifting ideas about what constitutes the urban, and urban natures. Participants are invited to imagine urban space as a site of environmental sustainability and to connect their bodies with the materiality of urban vegetal spaces. Echoing the words of environmental scholar Shiuhhuah Chou (2018) writing of her foraging as a Chinese women in a New York context, the Gatherings and walks remind us that cities are "productive and fertile", not just "impure", and "damaged", reconfiguring the city as a site of potential food and social justice for racialised bodies (Chou, 2018).

The events are designed for attendees to develop hopeful relations with urban nature and agriculture, encouraging minoritised groups such as people of colour, people with disabilities, white women to reimagine and make their own place in the city, in sensory, embodied and imaginative ways. In so doing they encourage the active, embodied participation of those who are Othered by the mainstream "food growing and environmental movement who may experience poor green space provision and not be able to leave the city easily" (MacGregor et al., 2019).

The Network activities can be understood through a lens of environmental justice in that they enable a "de-naturalisation" of who is seen as an expert in, and who is allowed to inhabit, enjoy, and learn about urban nature, green space, and environmentalism (Gray & Sheikh, 2018, 2021). Women of colour are treated as expert gardeners, with specialist understandings of Caribbean Somali and Bangladeshi veg and fruit, and creative gardening and sustainability practices. As scholars argue colonial histories and white environmentalism have excluded the expertise, knowledge and participation of racially minoritised people (Agyeman et al., 2016; Gray & Sheikh, 2018). WEN speaks to minoritised women and men and children as existing or potential eco-activists and ethical sustainable producers, challenging the normative bodily dispositions of white environmentalism (McClintock, 2018, p. 582).

Feminist scholars stress that urban public spaces are not universally accessible by gender, race, faith and class, constraining participation, feelings of safety and welcome and limiting roles due to racism and violence (Day, 1999). Indeed, walking as a practice is embedded within broader power geometries, social contexts, relations and structural differences including police governance of urban space, racialised laws, and racism (O'Neill & Roberts, 2019). Such "forms of segregation become ingrained into the material fabric of space and reflected in the practices and rhythms of everyday life", including the green spaces of TH (Robinson & McClelland, 2020, p. 655). As black feminist bell hooks argues, the street is full of "racial terror" for black people and not a routine or habitual space of encounter (Smith, 2016). As I indicated in my history of TH, racially minoritised women especially Muslim women wearing hijabs have been terrorised on the streets and sometimes in their own homes. The Gatherings and the walks build on racially minoritised groups own resistances by facilitating the occupation of exclusionary spaces and more equitable access to environmental spaces and resources.

The Gatherings and walks offer a multiracial and multi-embodied viscosity, which challenges the white viscosity of many alternative food initiatives (Ramírez, 2015). It's unusual to see a group of racialised women, disabled women and children walking in the streetscapes and white coded green spaces offered by the WEN events. The Gatherings encourage them to access different natural and food growing spaces and to inhabit the specific materialities of the vegetal spaces in embodied, sensory ways, consuming, producing and "troubling" space, and reconfiguring everyday practices of space and notions of food (Robinson & McClelland, 2020).

Troubling the everyday

As mentioned in the introduction, I want to hold onto these hopeful encounters but also flag some troubles which I do here by discussing how the walks don't refer to the colonial histories of the plants and green spaces in TH. Colonial histories and imperial formations continue to "haunt", and leave material and immaterial, situated and site-specific "traces" but often erased or marginalised in white media and histories as a kind of amnesia (Wemyss, 2008). In some ways, it's not surprising. Whilst some of the colonialities of the East End docklands and East India Company are known, the colonialities of nature, "imperial landscaping practices" and botany, and the East India Company's role in these, are much less well known, particularly in relation to British community food initiatives (Casid, 2005). The historic colonial "ecological networks" of edible and ornamental plants and the movement of seeds, saplings, seedlings and cuttings have been repressed (Gandy & Jasper, 2020; Czennia, 2021; Fowler, 2020; Ratinon & Ayre, 2021; Schiebinger & Swan, 2007). As Colonial historians Londa Schiebinger and Claudia Swan argue, "botany was 'big science' in the early modern world; it was also big business, enabled by and critical to Europe's burgeoning trade and colonialism" (2007, pp. 2–3).

Furthermore, scientific classifications of plants were related colonial hierarchies of race (Fowler, 2020; Ernstson, 2020). Imperial ideals of nature, plants and gardening were deployed to articulate notions of superiority, femininity and nationality (Hong, 2021). Plants in gardens and parks symbolised a material imperial power, subsidised by wealth from slavery, creating disastrous planetary level effects today (Fowler, 2020; Gandy & Jasper, 2020). For environmental scholar Henrick Ernstson, these histories "structure social relations and imaginaries," and condition how "urban nature can be conceived, engaged and known", an important point to consider for WEN and TH (2020, p. 73).

What if we were to critically engage some of these issues in community food initiatives? How might the walks be designed to highlight the colonial remains within urban ecology and urban environmental knowledge today? Plants shaped imperial landscapes, adorned gardens, cities and plantations but how are these embedded in TH? How might these reconfigure not only our cognitive understandings but also sensory, embodied, and visceral pleasures? As gardener Clare Ratinon and artist Sam Ayre write "horticulture is rarely seen as violent or associated with pillage and theft but how else do we think the plants from other parts of the world got to be here?" (2021, p. 5). The histories of routes which transported people, spices and sugar are documented in East London histories, but less is known about the botanic routes of plants, saplings and seeds (Ratinon & Ayre, 2021). As they continue, these practices "decimated landscapes, and eco-systems, fuelled plantations", and erased or exploited indigenous and gendered knowledges (Ratinon & Ayre, 2021).

Given WEN's connection ecofeminism, I want to reflect too on the gendered, classed, and colonial relations of imperial botany and trade. Of great significance is that colonialists learned from local women and midwives but erased their knowledge in the process of creating codified knowledge systems incorporated into European taxonomies in which they objectified plants and abstracted them from cultural practices (Fowler, 2020; Schiebinger, 2004). Historian Jiang Hong (2021) details how in the 19th-century European women consumed "trophies of colonial botany" in their everyday domestic practices: cooking in their kitchens, reading botanical books in their libraries, and decorating their houses and gardens with "exotic" plants. As wives in the colonies, they created Anglicised gardens, moving European plants to the colonies and botanising for their fathers, brothers, or husbands. Hong insists that their feminised practices endorsed imperialism and imperial notions of nature. At the same time, we can find histories of resistance and survival. Gendered knowledges were key to enslaved people's survival, including how enslaved women created food gardens as modes of sanctuary (Carney, 2021), and used plants as abortifacients to stop their children from becoming enslaved (Schiebinger, 2004).

These histories underline how the colonial underpins the everyday. As artists Sheila Ghelani and Sue Palmer write of the EIC and London, the colonial lies "in our country estates; our spice racks; our fabric patterns; in the foundations of the Lloyds Building; in the dock basin beside Canary Wharf" (Ghelani & Palmer, 2020, online). With this quote they stress the material, symbolic and something traces of colonial imperial histories in our bodies, the intimacies of the domestic sphere and public buildings and infrastructure. Of course, walks are limited in how they can address these issues, but walking has been used "to illuminate forgotten, ignored, or taken-for-granted features of the political, material, and cultural landscape" (Robinson, and McClelland, 2020, p. 654). Historians and public bodies are beginning to retell histories of gardens and stately homes, disrupting notions of domesticity, civility and sophistication with which they are often associated, and reminding people where the wealth came from and what it funded that will structure everydayness (Fowler, 2020; Smith, 2015, 2016).

Re-walking the walks

With this in mind, let's return to Stepney Green, the site of Stepney Farm, the place for the walk I describe and look for ghosts and traces. In this section, describe specific colonial garden, food and domestic histories of Stepney and how they intersect with the walk. The bigger histories of the colonial and imperial histories of the Docklands nearby are more well known, even if racially minoritised residents' connections are erased (Wemyss, 2008). My aim is to show how a walk could be retold to stay with the troubles of domestic and food related colonialism and imperialism by reconnecting to local sites.

Stepney was site of agricultural land and market gardens for the City of London in medieval times. As the EIC grew, Stepney became a site for the homes of global merchants, and industrial elites and their maritime or shipping-based business working in the colonies and slave trade. Stepney was close to the Thames and the City of London. East End of London historian, David Morris, writes that "in a short walk along the road it would have been possible to meet merchants trading with Hudson Bay, Greenland, Virginia, the Levant and Africa, but dominating discussion would have been the affairs of the East India Company" (Morris, 1986, p. 20).

The EIC was also central to imperial botany. Barbel Czennia explains that "institutional networks of merchants, military personnel and imperial administrators . . . especially those from the EIC created cultural and commercial exchanges of plants, gardeners, garden concepts and garden architecture, knowledge and aesthetics between Britain and the colonies" (2021, p. 93). TH was a site of important nurseries in the 18th and 19th centuries, key areas for cultivation of colonial exotic plants (Alcorn, 2020). Professional nurseries based in London met the growing demands for tropical and rare plants and their owners were critical intermediaries in the colonial exchange of plants in between scientific networks, commerce and publics, transforming plants from specimens into commodities (Alcorn, 2020; Czennia, 2021).

The East End of London was the site for an "arc" of such plant nurseries, extending from Hackney, to Clapton, the Lea Valley and Mile End, the latter being where most of the Gatherings and walks take place (Alcorn, 2020). Just up the road from the Farm was the site of James Gordon's Mile End world renowned nursery and glasshouses, founded in 1742. Gordon was a famous gardener and seed supplier trading in the colonial economies of plants and seeds, and with expertise in growing American plants from the newly invaded and colonised America, and plants from China. (Morris, 2000). Elite women and men in large numbers visited nurseries such as his to buy plants for their gardens and interiors, to display their wealth and connections to the imperial project and as a shopping day out, to be seen (Alcorn, 2020; Hong, 2021). Gordon also had links with Carl Linnaeus, the 18th-century Swedish botanist, whose classifications of humans underpins modern scientific racism, and the history of anthropology with devastating and far-reaching consequences including the dehumanisation of non-Europeans and justification for like slavery and indigenous genocide (Linneas society).

Early colonial horticulturalist plants such as Gordon and his peers sold are now the mainstay of parks in London and gardens. Gardening historian Christopher Tilley notes that English gardens "provided a kind of living material map of colonialism, exploration, and globalization" (2008, p. 227). He writes that many flowers presumed to be Ur-English came from all over the world. Such understandings challenge nativist and nationalist ecological taxonomies and potential our sensory encounters (Gandy, 2022).

From 1700 onwards, 789 British country estates with parks, gardens, hot houses, and "exotic" planting belonging to colonial merchants and the EIC were built or developed (Fowler, 2020). In Stepney, 17th- and 18th-century several houses were associated with the EIC. Number 37, a grand gated, four-storey mansion with garden in Stepney Green, built around 1694, the oldest house in Stepney, was owned by Dormer Sheppard, a slave owner and merchant (Ridge, 1998). He was involved in tobacco production and indentured labour in the new colonies in Virginia, USA, and in the Caribbean trading in 60 enslaved Africans whom he transported to his plantation in St Kitts in 1620. In 1707, in a terrible historic event, he advertised four times in London papers for the return of "a black Boy named Lewis, about 15 years old, in a Fustian Frock with Brass Buttons, Leather Breeches and blue stockings who has run away". Shockingly but in a commonplace move at that time, he *bequeathed* enslaved people to his son in his will, an awful reminder of the dehumanising violence of colonialism, and his son continued to trade in the Caribbean.

Like many of its neighbours, No. 37 remained the dwelling of EIC officials and shareholders until 1764. In 1714, Lady Mary Gayer, bought no 37 with wealth accumulated from imperial and colonial projects. She was the widow of General Sir John Gayer who had been East India Company's General Governor of what was called Bombay, and known today as Mumbai. Her initials, "MG", are visible on the gates. From 1757 to 1764, it was owned by Lawrence Sulivan, a Director and Chairman of the East India Company who controlled a vast commercial political colonial empire wreaking devastation on people, animals and ecologies.

Further up from 37, the farm we visited stands on the site of Worcester House, a moated noble Manor House which had its origins in the Medieval period 1450–1550. Thompson owned this house too in the later 1640s and 1650s, along with others in other parts of the United Kingdom. In 2011, Crossrail, Europe's largest construction project, started archaeological investigation at the site of the farm. Alongside the remains of buildings, archaeologists found remnants of foods in the ground of the Farmland. They dug up fruit stones and pips from figs, plums, cherries and apples, walnut and hazelnut shells, black pepper, and the remains of marrow and pumpkin. Some of these could have been grown in the orchards and gardens.

But what's important here are the intimate, domestic embodied connections to colonialism and the reach of the EIC. As historian. Ben Miller (2016) writes, these foods "allude to the remote reach of emergent [racial] capitalism." Indigenous peoples in North America grew pumpkins and the invasion of their lands lead to the arrival and later cultivation in the United Kingdom. In similar vein, but in India and Sumatra during the 16th-century black pepper, or "black gold", as it was known because of its expense, was one of the four major cargoes carried by the EIC (Ravindran, 2000). As Shannon Woodcock reminds us "food production was both the rationale for and the site of colonial expansion" (2016, p. 34). European cities have been "fed with the products of plantations and slavery", with today's agro-food businesses grown from colonial and masculinist roots (Ferrando et al., 2021, p. 62).

These histories in TH show how EIC, imperialism and colonialism impacted not only the docklands and civic architecture but also local domesticity and every days. Smith (2015, 2016) reminds us that it was partly through such everydays that notions of racial hierarchy were constructed. The houses and land remind of the concealed violence and amassed colonial wealth from empire.) Signs at Stepney Farm indicate that they intend to recreate the 18th-century manor house garden, and therein lies an opportunity to work with local residents to reconnect with these more intimate, domestic histories. They can stimulate new understandings about what's encoded in the landscape, plants, food and green spaces, opening up new interpretations of nativist ecologies and botanic taxonomies and the ground of the everyday. They underline the roots and routes of the interconnectedness of nature and human culture through history and their devastating effects for people and ecologies then and now (Damodaran, 2015; Gandy, 2022). They point to the historic and ongoing "geometries of power" that dictate who and what is where, and how they got there, and the colonial and contemporary connections between "here" and "over there" (Massey, 2005; Jeppesen, 2018).

Conclusion

In this chapter, I offer what the editors call a critical reparative approach to reflecting on the outreach activities of the Women's Environmental Network in TH in London, United Kingdom. In holding hopes of place-making, intercultural mixing, sharing expertise, and sensory encounters and embodiment – and troubles – of devastating colonial, imperial histories, racisms, exclusions, and inequalities – in tension through the "re-walking" structure of my chapter, I seek to find ways to write the critical reparative in line with the aims of this book. More specifically, I seek to show how everydayness, everyday environmentalism, hopes, and troubles are unevenly distributed by race, gender, and class. But as recent colonial histories of ecologies, gardens, and botany proliferate in the United Kingdom (Fowler, 2020; Ratinon & Ayre, 2021; Ware, 2021), as the Race, Empire and Education collective put it, teaching colonial and imperial histories in the United Kingdom can "repair the future" (REE, 2022).

I outline how WEN interpolates minoritised women, men, and children as activists, challenging white environmentalism, inviting participants, especially women of colour, to embody themselves as food producers, alternative consumers, and environmental subjectivities and to spend time in green spaces, engaging with diverse natures. WEN encourages minoritised people, especially women of colour and with refugee backgrounds, to share their transnational experiences and expertise and to find ways to produce communally situated, cosmopolitan environmentalisms in East London (Metcalf et al., 2012).

More specifically, I focused on specific forms of outreach education, understudied in research on community gardens, and showed how the walks and workshops offer sensual experiences of urban nature and modes of food production. They, and the multi-racial gardeners and allotment owners, produce new socio-ecological spaces of justice within the city, enabling attendees to become involved in green spaces through sight, hearing, touch, smell, and even taste and to notice cosmopolitan food growing, place making, and environmentalism, including the growing of crops from colonial elsewhere that have climate change benefits for the United Kingdom (Kell et al., 2018). Environmentally engaged walking is also a form of embodied activism challenging racial exclusions and white codes and viscosities. As scholars argue colonial histories and white environmentalism have excluded the expertise, knowledge, and participation of minoritised people, and their individual and collective activism (Agyeman et al., 2016; Gray & Sheikh, 2018). But WEN responds to such environmental racism through a feminist intersectional agenda, which encourages embodied knowledge sharing and learning, the valuing of minoritised food and environmental knowledges and practices and reclaiming urban nature and vegetal spaces for minoritised people.

Of course, environmental racism in a borough structured by racialised, gendered, and classed histories and hierarchies is far more invasive, intensive, and extensive than touched on here. But in relation to everyday environmentalism, activists and scholars can learn from colonial and imperial histories of food, places, and plants to trouble urban natures and white middle-class agricultural viscosities, understandings, and taxonomies. Minoritised members of the community offer an important source for alternative epistemologies about nature, gardening, and food, as seen in the WEN activities. Some attendees from ex-colonies have deep embodied and inter-generational understandings of colonial and imperial histories because of their own biographies.

Although in a minor key, the Gatherings and walks offer local people a small-scale, place based, and experimental approach consonant with the growth of new micro-political environmental movements, "rooted in everyday concerns" (MacGregor, 2021). At the same time, the chapter underlines how we cannot take the everyday – as space, temporality, or habit – for granted, especially in relation to colonial and imperial history, food, and nature, because in its various modalities, the everyday is often underpinned by colonial and imperial histories and legacies, often gendered, which activists and historians are increasingly reveal (BEN, Black2Nature; Fowler, 2020). Classed and gendered histories, racialisations, racism, and Islamophobia mean that some are excluded from the "condition of everydayness" whether that be in relation to food, sustainability, and growing practices, or access to urban and green space (Smith, 2015, p. 145).

References

Agyeman, J. (1990). Black people in a White landscape: Social and environmental justice. *Built Environment*, *16*(3), 232–236.

Ageyman, J. (2014). Entering cosmopolis: Crossingover, hybridity, conciliation, and the intercultural city ecosystem. *Minding Nature*, *7*, 2–16.

Agyeman, J., Schlosberg, D., Craven, L., & Matthews, C. (2016). Trends and directions in environmental justice: From inequity to everyday life, community, and just sustainabilities. *Annual Review of Environment and Resources*, *41*(1), 321–340.

Agyeman, J., & Spooner, R. (1997). Ethnicity and the rural environment. In P. Cloke & J. Little (Eds.), *Contested countryside cultures* (pp. 197–217). Routledge.

Alcorn, K. (2020). From specimens to commodities: The London nursery trade and the introduction of exotic plants in the early nineteenth century. *Historical Research*, 93(262), 715–733.

Alkon, A. H., & McCullen, C. G. (2011). Whiteness and farmers markets: Performances, perpetuations . . . contestations? *Antipode*, *43*(4), 937–959.

Aziz, A. (2021). Globalization, class, and immigration: An intersectional analysis of the new east end. *Sage Open*, *11*(1). https://journals.sagepub.com/doi/full/10.1177/21582440211003083

BEN. www.ben-network.org.uk/

Black2Nature. www.birdgirluk.com/black2nature/

Brahinsky, R., Sasser, J., & Minkoff-Zern, L. A. (2014). Race, space, and nature: An introduction and critique. *Antipode*, *46*(5), 1135–1152.

Brown, C., Husbands, C., & Woods, D. (2019). Transforming education for all: Tower hamlets and urban district education improvement. In *Innovations in educational change* (pp. 23–38). Springer.

Buckingham, S. (2005). Allotments and community gardens: A DIY approach to environmental sustainability. In S. Buckingham & K. Theobald (Eds.), *Local environmental sustainability* (pp. 195–212). Woodhead.

Byrne, J. (2012). When green is white: The cultural politics of race, nature and social exclusion in a Los Angeles Urban National Park. *Geoforum*, *43*(3), 595–611.

Carney, J. A. (2021). Subsistence in the plantationocene: Dooryard gardens, agrobiodiversity, and the subaltern economies of slavery. *The Journal of Peasant Studies*, 48(5), 1075–1099.

Casid, J. H. (2005). *Sowing empire: Landscape and colonization*. University of Minnesota Press.

Chou, S. (2018). Chinatown and beyond: Ava Chin, urban foraging, and a New American cityscape. *ISLE: Interdisciplinary Studies in Literature and Environment*, *25*(1), 5–24.

Curran, W. (2017). *Gender and gentrification*. Routledge.

Czennia, B. (2021). Green rubies from The Ganges: Eighteenth-century gardening as intercultural networking. In B. Czennia & G. Clingham (Eds.), *Oriental networks: Culture, commerce, and communication in the long eighteenth century*. Rutgers University Press.

Damodaran, V. (2015). East India company, famine and ecological conditions in eighteenth century Bengal. In V. Damodaran, A. Winterbottom, & A. Lester (Eds.), *East India company and the natural world*. Palgrave Studies in World Environmental History. Palgrave Macmillan.

Day, K. (1999). Embassies and sanctuaries: Women's experiences of race and fear in public space. *Environment and Planning D: Society and Space*, *17*(3), 307–328.

Domene, E., & Saurí, D. (2007). Urbanization and class-produced natures: Vegetable gardens in the Barcelona metropolitan region. *Geoforum*, 38(2), 287–298.

Driver, F., & Gilbert, D. (1998, February 1). Heart of empire? Landscape, space and performance in imperial London. *Environment and Planning D: Society and Space*, *16*(1), 11–28.

Engel-Di Mauro, S. (2018). Urban community gardens, commons, and social reproduction: Revisiting Silvia Federici's revolution at point zero. *Gender, Place & Culture*, *25*(9), 1379–1390.

Ernstson, H. (2020). Urban plants and colonial durabilities. In M. Gandy & S. Jasper (Eds.), *The botanical city* (pp. 71–81). Jovis Verlag GmbH.

Felski, R. (2000). *Doing time: Feminist theory and postmodern culture*. New York University Press.

Ferrando, T., Claeys, P., Diesner, D., Vivero-Pol, J. L., & Woods, D. (2021). Commons and commoning for a just agroecological transition: The importance of de-colonising and de-commodifying our food systems. In C. Tornaghi & M. Dehaene (Eds.), *Resourcing an agroecological urbanism* (pp. 61–84). Routledge.

Finney, C. (2014). *Black faces, White spaces: Reimagining the relationship of African Americans to the great outdoors*. UNC Press Books.

Forno, F., & Wahlen, S. (2022). Environmental activism and everyday life. In M. Grasso & M. Giugni (Eds.), *The Routledge handbook of environmental movements* (pp. 434–450). Routledge.

Fowler, C. (2020). *Green unpleasant land*. Peepal Tree Press.

Gandy, M. (2022). *Natura urbana: Ecological constellations in urban space*. The MIT Press.

Gandy, M., & Jasper, S. (2020). *The botanical city*. Jovis.

Ghelani, S., & Palmer, S. (2021). *Common salt*. LADA.

Ghelani, S., & Palmer, S. (2020). *On salt and tax*. http://feastjournal.co.uk/article/on-salt-and-tax/

GLA. (2015). '*Annual London Survey 2015*' available at https://data.london.gov.uk/dataset/annual-london-survey-2015

Gray, R., & Sheikh, S. (2018). The wretched earth: Botanical conflicts and artistic interventions introduction. *Third Text*, *32*(2–3), 163–175.

Gray, R., & Sheikh, S. (2021). The coloniality of planting. *Architectural Review*, *1478*, 14–17.

Guthman, J. (2008a). Bringing good food to others: Investigating the subjects of alternative food practice. *Cultural Geographies*, *15*(4), 431–447.

Guthman, J. (2008b). "If they only knew": Color blindness and universalism in California alternative food institutions. *The Professional Geographer*, *60*(3), 387–397.

Hall, S. M., Pottinger, L., Blake, M., Mills, S., Reynolds, C., & Wrieden, W. (2020). Food for thought?: Material methods for exploring food and cooking. In H. Holmes & S. M. Hall (Eds.), *Mundane methods: Innovative ways to research the everyday*. Manchester University Press.

Hickcox, A. (2018). White environmental subjectivity and the politics of belonging. *Social & Cultural Geography*, *19*(4), 496–519.

Hong, J. (2021). Angel in the house, angel in the scientific empire: Women and colonial botany during the eighteenth and nineteenth centuries. *Notes and Records*, *75*(3), 415–438.

Jarosz, L. (2011). Nourishing women: Toward a feminist political ecology of community supported agriculture in the United States. *Gender, Place & Culture*, *18*(3), 307–326.

Jeppesen, C. (2018). Growing up in a company town: The East India company presence in South Hertfordshire. In M. Finn & K. Smith (Eds.), *The East India company at home, 1757–1857*. UCL Press.

Johnson, L. (2018). Becoming "enchanted" in agro-food spaces: Engaging relational frameworks and photo elicitation with farm tour experiences. *Gender, Place & Culture*, *25*(11), 1646–1671.

Kell, S., Rosenfeld, A., Cunningham, S., Dobbie, S., & Maxted, N. (2018). The benefits of exotic food crops cultivated by small-scale growers in the UK. *Renewable Agriculture and Food Systems, 33*(6), 569–584.

LBTH. (2013). '*Ethnicity in Tower Hamlets Analysis of 2011 Census data*' available at https://www.towerhamlets.gov.uk/Documents/Borough_statistics/Ward_profiles/Census-2011/RB-Census2011-Ethnicity-2013-01.pdf

Le Grand, E. (2020). Moralization and classification struggles over gentrification and the hipster figure in austerity Britain. *Journal of Urban Affairs*, 1–16.

Linnaess Society (2021). *Linnaeus and race*. Retrieved May 22, 2022, from www.linnean.org/learning/who-was-linnaeus/linnaeus-and-race

Listerborn, C. (2020). Gender and urban neoliberalization. In A. Datta, P. Hopkins, L. Johnston, E. Olson, & J. M. Silva (Eds.), *Routledge handbook of gender and feminist geographies*. Routledge.

MacGregor, S. (2021). Making matter great again? Ecofeminism, new materialism and the everyday turn in environmental politics. *Environmental Politics*, 30(1–2), 41–60, https://doi.org/10.1080/09644016.2020.1846954

MacGregor, S., Walker, C., & Katz-Gerro, T. (2019). It's what I've always done: Continuity and change in the household sustainability practices of Somali immigrants in the UK. *Geoforum, 107*, 143–153.

Martin, M. A. (2019). Digging through urban agriculture with feminist theoretical implements. *Canadian Food Studies/La Revue Canadienne Des Études Sur L'alimentation*, 6(3), 88–107.

Massey, D. (2005). *For space*. Sage Publications.

Matless, D. (2016). *Landscape and Englishness* (2nd expanded ed.). Reaktion Books.

McClintock, N. (2018). Urban agriculture, racial capitalism, and resistance in the settler-colonial city. *Geography Compass, 12*(6), 1–16.

Mell, I., & Whitten, M. (2021). Access to nature in a post Covid-19 world: Opportunities for green infrastructure financing, distribution and equitability in urban planning. *International Journal of Environmental Research and Public Health*, 18(4), 1527.

Metcalf, K., Minnear, J., Kleinert, T., & Tedder, V. (2012). Community food growing and the role of women in the alternative economy in tower hamlets. *Local Economy, 27*(8), 877–882.

Miller, B. (2016). *Stepney Green: Archaeologists reveal shoes, goblets and more of ancient EastEnders*. www.culture24.org.uk/history-and-heritage/archaeology/art547832-stepney-green-crossrail-archaeology-london-mansion

Mollett, S. (2017). Gender's critical edge: Feminist political ecology, postcolonial intersectionality, and the coupling of race and gender. In S. Macgregor (Ed.), *The Routledge handbook of gender and environment*. Routledge.

Morris, D. (1986). Mile end old town and the East India company. *East London Record*, 9, 20–28.

Morris, D. (2000). James Gordon, mile end's famous nursery man. *Transactions of the London & Middlesex Archaeological Society*, 51, 183–189.

O'Neill, M., & Roberts, B. (2019). *Walking methods: Research on the move*. Routledge.

Panelli, R., Hubbard, P., Coombes, B., & Suchet-Pearson, S. (2009). De-centring White ruralities: Ethnic diversity, racialisation and indigenous countrysides. *Journal of Rural Studies*, 25, 355–364.

Pitt, H. (2015). On showing and being shown plants – a guide to methods for more-than-human geography: On showing and being shown plants. *Area*, 47(1), 48–55. https://doi.org/10.1111/Area.12145

Rahman, M., & Billah, M. (2021). Case study 3: The tragic death of Altab Ali and the beginning of confrontation against racism and fascism. In D. Hack-Polay, A. B. Mahmoud, A. Rydzik, M. Rahman, P. A. Igwe, & G. Bosworth (Eds.), *Migration practice as creative practice*. Emerald Publishing Limited.
Ramírez, M. M. (2015). The elusive inclusive: Black food geographies and racialized food spaces. *Antipode*, *47*(3), 748–769.
Ratinon, C., & Ayre, S. (2021). *Horticultural appropriation: Why horticulture needs decolonising*. Rough Trade Books and Garden Museum.
Ravindran, P. N. (Ed.). (2000). *Black pepper: Piper nigrum*. CRC Press.
REE (2022). *Race, empire and education collective*. REE.
Reynolds, K., & Cohen, N. (2016). *Beyond the kale: Urban agriculture and social justice activism in New York City* (Vol. 28). University of Georgia Press.
Ridge, T. (1998). *Central Stepney history walk*. Central Stepney Regeneration Board.
Rifkin, S. (2018). *Cultivating community: Towards a Black women-centered alternative food politic*. https://openscholarship.wustl.edu/wuurd_vol13/177
Rishbeth, C., & Birch, J. (2020). Urban nature and transnational lives. *Population, Space and Place*, E2416.
Robinson, J., & McClelland, A (2020). Troubling places: Walking the "troubling remnants" of post-conflict space. *Area*, *52*(3), 654–662.
Sankey, K. (2015). Peasant revolts in an era of globalization: Bringing political economy back in. *Critical Sociology*, *41*(7–8), 1199–1209. https://doi.org/10.1177/0896920515591297
Schiebinger, L. (2004). *Plants and empire: Colonial bioprospecting in the Atlantic world*. Harvard University Press.
Schiebinger, L., & Swan, C. (Eds.). (2007). *Colonial botany: Science, commerce, and politics in the early modern world*. University of Pennsylvania Press.
Sharma, M. (2020). Postage stamps, war memory, and commemoration: A case study of the Bangladesh liberation war of 1971. In F. Jacob (Ed.), *War and semiotics*. Routledge.
Slocum, R. (2008). Thinking race through corporeal feminist theory: Divisions and intimacies at the Minneapolis farmers' market. *Social & Cultural Geography*, 9(8), 849–869.
Smith, A. (2016). *Racism and everyday life: Social theory, history and "race."* Palgrave Pivot.
Smith, A. (2015). Rethinking the "everyday" in "ethnicity and everyday life." *Ethnic and Racial Studies*, 38(7), 1137–1151.
Smith, D. (1987). *The everyday world as problematic: A feminist sociology*. University of Toronto Press.
Thompson, C., & Macgregor, S. (2017). The death of nature: Foundations of ecological feminist thought. In S. Macgregor (Ed.), *The Routledge handbook of gender and environment* (pp. 43–54). Routledge.
Tilley, C. (2008). From the English cottage garden to the Swedish allotment: Banal nationalism and the concept of the garden. *Home Cultures*, 5(2), 219–249.
Tolia-Kelly, D. P. (2004). Landscape, race and memory: Biographical mapping of the routes of British Asian landscape values. *Landscape Research*, *29*(3), 277–292.
Vehviläinen, M. (2017). Practices of modest recuperation: Food, situated knowledge, and the politics of respect. *International Journal of Gender, Science and Technology*, 9(2).
Walter, P. (2013). Theorising community gardens as pedagogical sites in the food movement. *Environmental Education Research*, 19(4), 521–539.
Walton, S. (2021). *Everybody needs beauty: In search of the nature cure*. Bloomsbury.

Ware, V. (2021). *Return of a native: Learning from the land*. Repeater Books.
Wemyss, G. (2008). White memories, White belonging: Competing colonial anniversaries in "postcolonial" East London. *Sociological Research Online*, *13*(5), 50–67.
Woodcock, S. (2016). Biting the hand that feeds: Australian cuisine and aboriginal sovereignty in the great sandy strait. *Feminist Review*, *114*(1), 33–47.

10 Eating (with) the other

Staging hope and trouble through culinary conviviality

Oona Morrow

Introduction

In different cities across the world, we have seen a sharp rise in community food initiatives aimed at spurring social change and inclusion through practices of cooking and eating together (Davies et al., 2017; Davies, 2019; Edwards, 2021; Marovelli, 2018; Smith & Harvey, 2021). Each of these social innovations in commensality and conviviality is responding to unique characteristics, tensions, and conflicts in their localities, the needs of the populations they serve, and broader geopolitical forces that affect the dynamics of mobility and migration to these places.

In this chapter, I examine two community cooking initiatives that seek to valorise the culinary heritage, knowledge, and foodways of migrants in New York City and Berlin. In New York City, the *League of Kitchens* places a special emphasis on valorising the skills of immigrant women by staging paid in-home cooking workshops. In Berlin, *Über den Tellerrand* offers paid workshops and community cooking events at Kitchen Hub that provide a setting for refugees and asylum seekers to share their food culture with locals. In different ways, these community food initiatives use food knowledge, cooking, and eating to address the troubles of racism and xenophobia. They offer migrants and locals the hope of cross-cultural friendships and social networks, whilst generating employment opportunities for people who are marginalised from the labour market and devalued in the food service industry. The relative success of these initiatives depends on the emotional and performative labour of cooks who stage authenticity and identity and on the desires of an affluent group of consumers and food enthusiasts who are willing to pay a premium for these experiences. This chapter examines how cooks and eaters within these initiatives navigate the power asymmetries of this transaction in their shared desire for transformative culinary encounters.

Eating difference: conviviality, commensality, and intercultural encounters

In popular media and academic literature, the idea that "food connects" and helps people understand, empathise, and overcome difference is widely celebrated. "We all eat," or so the saying goes. This hopeful appraisal of culinary conviviality relies

DOI: 10.4324/9781003195085-13

on the belief that the practice of commensality (Dunbar, 2017), the act of eating together, can bring everyone to the same level in spite of existing power inequalities. This is part of a broader trend in community food initiatives that emphasise the power of food to solve a variety of troubles ranging from health inequalities to poverty, environmental awareness, and social exclusion (Hayes-Conroy & Hayes Conroy, 2013).

A growing body of food studies scholarship examines food as a site of multi-cultural encounter, commensality, and conviviality (Wise, 2012; Molz, 2007; Heldke, 2005; Johnston & Bauman, 2015). According to Wise and Noble (2016), conviviality, or the capacity to live together, has become "the latest groovy thing" in social theory. Investigations of conviviality address fundamental sociological concerns related to gifting, reciprocity, and community formation and contestation. Rather than viewing conviviality as "happy togetherness" they argue "that conviviality is useful only if it is understood in a very specific way; a way that includes potential ambivalence at the heart of the everydayness of living together" (2016, p. 425). While conviviality is often observed in the moment, as a vibe, or an assumed outcome of commensality, Phull et al. (2015) go a step further to analyse the conditions that allow culinary conviviality, or the pleasure of eating together, to happen. They are also mindful of the social function of eating together, which can include segregation, division, and group formation. "The way we eat and whom we eat with is symbolic of the way society divides itself through class, kinship, age or occupation and may result in social exclusion for those not part of commensal circles" (Kerner et al., 2015; Phull et al., 2015, p. 979). In other words, conviviality doesn't just "happen" when people eat together, rather "a group needs to 'play by the rules' of sociable interactions to construct a pleasant eating event. It is in this way that conviviality may differ from commensality" (Phull et al., 2015, p. 979). From this perspective, we can begin to view culinary conviviality, the pleasurable act of eating together, as the outcome of a number of cultural, emotional, and social performances that are situated and staged in kitchens and around tables. Following Wise and Noble (2016), it is also important to consider that culinary conviviality may not always be (equally) pleasurable and can come with conflict and tension in the negotiation of differences and power inequalities.

The extent to which such intercultural culinary convivialities are transformative (and for whom) or fall into the trap of culinary colonialism and exoticism is widely debated (Alkon & Grosglik, 2021; Heldke, 2013; Wise, 2012). The black feminist scholar bell hooks coined the term "eating the Other" to describe the ways in which Otherness has been commodified as a "seasoning that can liven up the dull dish that is mainstream white culture" (1992, p. 366). Encounters with the Other can serve to reinforce "white racism, imperialism, and sexist domination" for consumers and eaters who assert their power and privilege while enhancing their cultural and culinary capital through acts of "courageous consumption" (hooks, 1992, pp. 378–379). This dynamic is captured vividly by food journalist Soleil Ho (2014) and comic artist Shing Yin Khor (2014), who both illustrate the manifold ways in which adventure seeking "Foodies" (see, for example, Johnston and Bauman 2015) engage in acts of cultural appropriation in their quest for authenticity, identity, and status through consuming "ethnic" cuisine and converting this to cultural capital. In a critical self-reflection on her own penchant for "ethnic" foods from economically dominated cultures, philosopher Lisa Heldke (2013) recognises an attitude she names "cultural food colonialism." While genuinely respectful and curious about other food cultures, she also observes, "I was motivated by a deep desire to have contact with – to somehow own an experience – of an exotic Other as a way of making myself more interesting" (Heldke, 2013, p. 395).

A less extractive account of eating across differences is captured by Alkon and Grosglik (2021) in their analysis of Anthony Bourdain's food travel television show. In contrast to the now familiar trope of "eating *the* Other" (hooks, 1992), the authors also identify "eating *with* the Other" which they describe as "a warm and respectful entree into the everyday realities of racial, ethnic, and immigrant communities" (2021, p. 10). In these respectful encounters, "hosts are eating with the Other as relative equals, sharing food, conversation, and commensality" (2021, p. 6). However, considering the power geometries within which such encounters take place, they can also "operate as a slippery slope towards trends of eating *the* other, when discourses and engagements with the foods of the other people are revealed as a means of reinforcing racial and class hegemonies" (2021, p. 11). However, as several authors (Wise, 2012; Heldke, 2005) make clear, there is very little that separates eating *with* the Other from eating *the* Other, and variables such as setting, performance, food, and general vibe can tip the balance and trigger a range of positive and negative emotions in participants.

The emotions of desire and discomfort are key to understanding respectful (eating *with* the Other) and disrespectful (eating *the* Other) intercultural culinary encounters and can shape how transformative these encounters are (and for whom). Like the categories of hope and trouble, the emotions of desire and discomfort are not entirely separate, and may co-exist in the same experience for different actors and within individual actors. As hooks reminds us, desire in itself is not problematic and can even be transformative, "Acknowledging ways the desire for pleasure, and that includes erotic longings, informs our politics, our understanding of difference, we may know better how desire disrupts, subverts, or makes resistance possible" (1992, p. 380). Discomfort on the other hand tends to be actively avoided in culinary encounters across difference, in favour of a normative form of conviviality as "happy togetherness." As Wise observes in her research on everyday multiculturalism, there is an assumed positivity to such encounters, "initiatives often simply assume that eating the food of the 'Other' in intercultural situations will have positive outcomes for race and interethnic relations." (2012, p. 84). In a similar vein, Alkon and Grosglik observe an emphasis on pleasure, positive feelings, and comfort to render the Other and their foods, palatable and comfortable for presumed white viewers and eaters. They argue that "connections across difference *require* discomfort if they are to challenge hierarchical social relations" (2021, p. 10). Wise elegantly weaves these dimensions of discomfort and desire together when she writes, "[food] can be a the subject of disgust and desire, mediating cultural difference in multicultural settings . . . because it is at once everyday, deeply embodied, and yet so symbolic of difference" (2012, p. 83).

However, Wise also makes clear that it is much more than food that mediates cultural difference. In response to the significant but often oversimplified "role of food in constructing, reconstructing and mediating cultural differences" (2012, p. 85), Wise asks, "under what conditions do experiences of 'otherness' through food make cosmopolitans or contribute to positive relationships across difference?" (2012, p. 85). She finds that it matters deeply who is eating with whom, where, and in what kind of setting. This chapter seeks to build on Wise's question, by exploring how community food initiatives are creating settings for convivial intercultural culinary encounters, the types of performances that constitute these encounters, and the role of setting, performance, desire, and discomfort in negotiating the power asymmetries of eating with others.

The dynamics of hope and trouble are evident in these different readings of commensality, conviviality, and intercultural culinary encounters. They can be found in the hopeful positivity that is often attributed to eating *with* the Other, as well as the ever-present dynamic of eating with *the* Other and the troubles of exoticism, colonialism, and white supremacy that feed this dynamic. On a more micro level there is great potential to explore hope and trouble in performances and experiences of intercultural culinary conviviality, by attending to emotions linked to hope and trouble, and following the flows of desire and discomfort that can be triggered and shared – in acts of cooking and eating together. This chapter examines these dynamics in two community food initiatives dedicated to staging intercultural encounters through cooking and eating together.

Background and methods

This chapter is based on ethnographic research conducted in Berlin and New York in the winter and summer of 2017. The research was conducted as part of a larger EU project on the sustainability potential of urban food sharing (SHARECITY, n.d., Davies, 2019). Case studies were developed in each city to capture different types of food sharing including communal growing, surplus food redistribution, cooking and eating together, and multi-functionality. In our scoping study (Davies et al., 2017) we encountered a diversity of initiatives that stage culinary conviviality through cooking and eating together, including informal and non-profit initiatives that cook and share surplus food (e.g. Food not Bombs, KuFA -Kitchen for All, Disco soup), as well as for-profit social dining platforms (Eatwith). We selected *Über den Tellerrand* (Berlin) and *League of Kitchens* (NYC) as comparative case studies because they both used cooking and eating together to respond to dynamics of migration in their respective cities.

In the wake of the Syrian refugee crisis in Berlin, activism and social entrepreneurship grew up around the needs of people who had been forced from their homes and were living in temporary shelters (often without kitchens), while facing severe restrictions on their mobility in Germany and Europe. This took place in a broader political and cultural climate where refugees experienced significant social isolation, dehumanisation, and outright racism and xenophobia from locals and right-wing political parties. This gave rise to protests across Berlin, culminating in the 2012–2014 occupation of Oranienplatz (a public square in Kreuzberg), where refugees, asylum seekers, local activists, and students built a protest camp. Students brought cooking equipment, cutlery, and food to set up convivial pop-up kitchens. Cooking and eating together at the camp gave rise to a number of spontaneous intercultural culinary encounters. Locals were inspired by these encounters and went on to found a number of refugee focused food initiatives in Berlin, including "*Über den Tellerrand*" (n.d.).

"*Über den Tellerrand*" (ÜDTR) operates out of a community cooking space called Kitchen Hub on a leafy residential street in the well-kept Schöneberg neighbourhood. At the time of the research, the staff and volunteers were primarily white German women in their 20s and 30s. The profile of the staff has intentionally become much more diverse since the research was completed as the organisation has grown. Organisationally ÜDTR is split between being a business (Gmbh), which offers paid cooking workshops, and a non-profit foundation (Verein), which can apply for funding to run a variety of community programmes that facilitate the sharing of food, knowledge, and skills for friendship building through community food events, language exchange, buddy programmes, beekeeping, gardening, and cultural events. All of the profits from the business are invested in the community programmes of the non-profit. The organisation also produces cookbooks with authentic regional recipes *Rezepte für ein besseres Wir* and fusion recipes *Eine Prise Heimat*, facilitates field trips and excursions, and provides informal job placement and training. Beyond Berlin, ÜDTR has launched Kitchen on the Run, a mobile community kitchen that travels across Europe, especially to places where there are tensions between refugees and locals. And there are now satellite ÜDTR projects in more than 37 cities, mostly in Germany but also in Colombia and the Czech Republic.

During the winter of 2017, I conducted participant observation at both paid cooking workshops led by refugee chefs and free community food events including: "50 Plates of . . ." a community cooking event and potluck, language cafes where migrants and locals practice a foreign language together, and cultural events and excursions. I conducted five semi-structured interviews with volunteers and staff, all of whom were white German women. I also met regularly with a group of male participants from Syria. They were not interviewed but provided significant insight into the user experience of ÜDTR. Chefs and participants tended to be male and reflected the broader demographics of the refugee population, while staff and volunteers tended to be white and female. Refugee women were less visible. There were a few paid female chefs, and I attended one workshop led by a woman from Afghanistan. In recognition of the mobility constraints that many women faced related to caring responsibilities as well as cultural taboos on going out alone or mixing in public, ÜDTR also facilitated day time women's only cooking events run by a group of Arabic refugees. These events were very important for women to socialise with one another outside of shelters, and importantly, they allowed them access to kitchen space to cook familiar foods that they brought back to their kitchen-less shelters to share. The community programmes have further expanded and diversified since this research was completed.

In the summer of 2017, as part of the SHARECITY project, I began a similar course of ethnographic fieldwork in New York City, and selected a comparative case for cooking and eating together that was also geared towards bringing different kinds of people together around the kitchen table in convivial intercultural culinary encounters. At the time of the research, New York did not have any refugee focused food sharing initiatives, which reflects the fact that the United States has accepted far fewer refugees than Europe, and has tended to settle them in remote locations. However, the troubles of racism, xenophobia, and social exclusion are prevalent. New York City is one of the most ethnically diverse cities in the United States, and is home to 3.1 million immigrants, and an estimated half-million undocumented immigrants (NYC Mayor's Office of Immigrant Affairs, 2018). Although they may not claim refugee status, many immigrants are also economic refugees. These immigrants are the backbone of New York's restaurant industry, working as chefs, dishwashers, busboys, waiters, bartenders, and food delivery people. For undocumented people, their immigration status makes them vulnerable to extremely exploitative working conditions, and the ever-present threat of deportation. Even though New Yorkers eat meals prepared by immigrants every day, they may not know anything about their lives and cultures, and their culinary knowledge is rarely valued. "Ethnic" restaurants are expected to be cheap, and immigrant chefs are often marginalised from the world of fine dining until their flavours and recipes are appropriated and domesticated by white chefs (Ray, 2017). Historical and everyday forms of racism and colonialism reinforce these dynamics.

In 2014, the social practice artist Lisa Gross founded the "*League of Kitchens*" (LoK), with the aim of staging intercultural encounters between locals and immigrants by cooking and eating together, documenting the culinary knowledge of immigrant home cooks, and valorising this knowledge through home cooking workshops. League of Kitchens is an intentional pun on League of Nations, an international diplomacy body that preceded the UN. It signals a hope in the power of food to bring people together, facilitate intercultural culinary encounters that bridge cultural differences, and nurture a greater appreciation for the knowledge and skills that immigrants bring to our foodscape. Perhaps because of the focus on providing a platform for non-professional home cooks, all the chefs are women. This is in sharp contrast to the world of professional chefs, which is dominated by men. The gender dimensions of domestic food work are celebrated on the League of Kitchens website, which offers "meals by grandma" and boasts "a culinary dream-team of women from around the world who will welcome you into their homes, teach you their family recipes, and inspire you with their personal stories" (League of Kitchens, 2022). Cultural exchange is at the heart of LoK home cooking workshops, which offer "meaningful connection and social interaction, cultural engagement and exchange, culinary learning and discovery, and exceptional eating and drinking. Through this experience, the League of Kitchens seeks to build cross-cultural connection and understanding, to increase access to traditional cooking knowledge, and to provide meaningful, well-paid employment and training for immigrants" (League of Kitchens, 2022). As a community food initiative, LoK is accountable to the needs of the close-knit community of chefs who make this project possible. These chefs are also part of multiple local and global communities that they invite guests to explore.

League of Kitchens is more than a platform for connecting home cooks with curious eaters: they also provide training and support to help women share their stories and skills. At the time of research, a staff anthropologist conducted interviews with each home chef to learn about her life, culture, ancestors, the place she was from, her neighbourhood, and foodscape. This interview is the basis for a highly personalised recipe booklet that is designed for each home chef and workshop. A professional recipe creator also worked closely with each chef to follow her kitchen and shopping routines and codify tacit and embodied knowledge – a pinch, a handful, etc., into standard measurements. During the course of my research, I conducted participant observation at two paid home cooking workshops with chefs in Brooklyn and Queens and performed interviews with two chefs and two staff members, all of whom were women of different cultural and ethnic backgrounds. During the workshops, I had informal chats with participants, but follow-up interviews were not possible.

During participant observation and in subsequent field notes, I paid close attention to the feeling and "vibe" of the event, as well as the different emotions participants and chefs seemed to be exhibiting and sharing, through tears, laughter, gustatory sounds, faces of disgust, and other forms of body language. I also observed my own emotional and affective experiences in these spaces while handling food, chopping, tasting, and eating with others. However, it was rare that these feelings were verbalised in interviews. This points to a more general challenge in conducting qualitative research using visceral methods (Hayes-Conroy, 2010). All interviews and field notes were transcribed and coded in NVivo with a set of pre-determined codes related to goals and motivations, rules and regulations, power dynamics, etc. A second round of coding was conducted for this book chapter to examine the themes in the literature related to eating *the* other vs. eating *with* the Other, desire and discomfort, the setting, and the performance.

Findings: staging intercultural culinary conviviality

In each initiative I observed elements of hope and trouble, which manifested around the emotions of desire and discomfort in performances of eating (with) the Other. Both participants and chefs were motivated by the hope that intercultural culinary encounters could result in respectful and convivial experiences of eating *with* the Other. However, other hopes and desires were also present, and could easily become a source of trouble and discomfort.

In paid cooking workshops, consumers often expressed a desire to acquire new knowledge and understanding. Sometimes this desire for knowledge extended beyond how to cook authentic recipes, and crossed a boundary into the acquisition of more intimate knowledge about a person's life and migration history. This could result in tensions for chefs when paying participants pressed them for details about fleeing violence and conflict, which could trigger painful memories of trauma. At the same time, migrant chefs and participants at community events sometimes expressed desire for more than encounters – for real friendships, deep connections, and social networks that could help them thrive. In this way we can also begin to see hope and trouble as more intertwined, dashed hopes and desires can also reveal existing power inequalities and cause troubles and discomfort.

Staging culinary conviviality demands significant emotional labour from chefs, whose performances aspire towards "happy togetherness," while protecting guests from experiencing discomfort. For example, in both of the homes I visited for LoK cooking workshops, chefs took great effort to treat guests as family. Sitting them down at the table for a homemade snack and a chat as soon as they arrived, before getting down to the business of cooking together. They also went to great lengths to make their homes inviting and cosy, even if that meant displacing other family members who could interrupt the workshop or change the vibe. The conversations we had were always rich and full of detail, stories, and jokes, but we were also careful to avoid uncomfortable topics – like the causes of their migration, US immigration and economic policies, fears of deportation, and racism against immigrants.

In both organisations, I observed that attendees at paid cooking workshops were mostly women and exclusively white native-born Germans (Berlin) or Americans (New York). Workshops took place in the afternoon and evening, with fees ranging from 75 Euro to 125 USD. Attending the cooking workshops requires considerable time and economic resources. LoK home workshops were limited to about 5 people, while ÜDTR Kitchen Hub workshops could accommodate 14. The home workshops did not serve alcohol, while the Kitchen Hub workshops did. Some participants had received the course as a gift, others were using it as a shared gift for family bonding, or as a gift to themselves. Participants were people who enjoyed food and cooking and were curious about another culture. Besides curiosity and pleasure, participants were motivated by the feeling of "doing good" by supporting women immigrants in New York and refugees in Berlin.

To better understand how the dynamics of hope and trouble manifest in cooking and eating together at LoK and ÜDTR, I examine how the different emotions are triggered by particular performances and settings. Comparing the public to the private kitchen, I show how each space has been designed to support particular performances of culinary conviviality. In this section, I unpack what is happening in these encounters, the setting of the stage, the possible performances and scripts, and the experiences of desire and discomfort that may result. This reveals a number of troubles in culinary conviviality, including the risk of eating *the* Other, provoking discomfort in traumatised bodies, and accommodating the conflicting desires of guests for both adventure, exploration, safety, and familiarity.

Settings

Following Wise's (2012) argument that the "social settings in which food is consumed cross-culturally matter immensely. The consumption of food always needs to be understood in relation to the settings in which it is consumed" (2012, p. 107). I begin by examining the different settings in which performances of intercultural culinary conviviality are staged, how these stages are designed, created, and shaped by participants, and the different performances, feelings, and experiences they support.

In LoK, the stage is the family home and kitchen. These are spaces that are usually private, personal, and hidden from the public behind closed doors. They are also spaces where care, rest, and social reproduction take place, and in New York, they can often feel tiny and cramped. The stage at LoK is both spatially and socially intimate. Setting the stage and getting it ready for five strangers requires a lot of work, cleaning, and negotiating with other family members for space as well as childcare.

At ÜDTR, the stage is Kitchen Hub, a spacious and sunny community space on the ground floor with large street facing windows on two sides. It is highly visible, and the activities inside are on display to the passers-by. Kitchen Hub is located in Schöneberg, in a corner property that has been vacant for more than ten years. It brings life and activity to a quiet and leafy residential street. The interior of Kitchen Hub was meticulously designed in collaboration with architecture students from Technical University Berlin, who have created flexible working, cooking, and eating spaces out of wood and metal. Most of the furnishings have multiple uses and a modular design. One wall is lined with shelves that hold cookbooks and a few austere jars of spices. In the back there is an actual kitchen, with a refrigerator, and sink for washing up, and space for food storage. The back kitchen is where real messes can happen. The front kitchen consists of tidy islands with a sink, cooktop, cutting block, and oven. To me, the space feels very neutral, even cold. However, staff and volunteers describe the space as both neutral and warm and inviting, and they focus on the wood as a source of warmth. Decisions have been made in the design to create a functional and neutral setting that can accommodate different activities, uses, and culinary identities.

> it's just a neutral place where everyone comes together. . . . But not too neutral or too sterile either, it should still be a place where you can come in and feel comfortable, that's important.
>
> I mean it's a lot of wood, wood is already warm. . . . It's so warm, it kind of feels warm, it's nicely decorated. Different spice jars that are there with Arabic spices and then again German spices, we have books there, plants, lots of them . . . a cozy seating area.
>
> (Staff, UDTR)

Participants find the space dignified, and much more posh feeling than a rundown community centre, shelter, or apartment kitchen. Staff were concerned with creating a "positive atmosphere" and providing a reprieve from the depressing, ugly, and stressful government and administrative spaces that refugees and asylum seekers must navigate to meet their needs. Kitchen Hub also reflects a certain cosmopolitan-hipster class habitus that is most at home in minimalist mid-century modern design. The space is not personal or intimate; it has been designed for the comfort of a diversity of users. It manages to be familiar to a class of paying guests who can feel like they are at a hip coffee bar, inviting and dignified to participants, and useful for a variety of functions – including community cooking.

The intentional and vernacular design of these spaces supports different kinds of performances of cooking and eating together. One important difference between the home kitchen and the community kitchen is the degree of autonomy and power chefs have over the design of the space, and the different roles and performances these stages support. At LoK, home chefs have complete control over their space, they are cooking in their kitchen, where they are the matriarch, even if this comes with some stress and inconvenience.

> [W]hen instructors are teaching in their home this is their space that they are welcoming people into and it just immediately sets up this feeling of hey, they're the host, the teacher, the expert, and they feel comfortable there, cooking there and being there and teaching there.
>
> (Staff, League of Kitchens)

At home surrounded by familiar objects, appliances, kitchen tools, and family photos, it is easy for chefs to slip into the familiar roles of hospitality and domesticity. The charming maternal host shows generosity but also power by serving her guests more food than they can possibly eat. The kitchen boss, who doesn't feel shy about delegating tasks, and making sure that you know how to do things "her way." This domestic stage at LoK is also disciplining for paying customers, who behave as careful and courteous guests rather than entitled customers. At the end of the cooking workshop, guests and the chef sit down at the dining table, where additional family members might join, and share a fantastic meal that they created together. The chef is still very much in charge of staging the meal; she ensures that the table is set up correctly, and the food is served and plated in a way that is beautiful to her. The setting is intimate and elegant, and by the end, guests are urging the chef to sit down, relax, and begin sharing responsibility for hospitality. In these moments of warmth, sharing, comfort, and respect, eating *with* the Other is temporarily achieved, and the Other is no longer an exotic stranger but an intimate family member or new friend.

At ÜDTR, the Kitchen Hub has been thoughtfully designed to be neutral and accommodating to a diversity of users and uses, but there is little room for chefs to make the space their own. They occupy the space as guests rather than residents, and they are on display. One volunteer observed that this can counter the aim of putting refugees and locals on equal footing.

> Put on exhibition! Like in the zoo! It's not supposed to be that way.
>
> (volunteer, ÜDTR)

For the evening, chefs can fill the space with music and smells from home, but they will always leave behind a clean slate for the next user. Within this space, refugees are on their feet and take on the professionalised roles of chef and expert. Paying customers sit around a big table and act more like customers than guests. They are not shy about "getting what they paid for." Guests tend to socialise more with each other over a few glasses of wine, while chopping ingredients, while the chef fades into the background and does the more serious cooking. The consumption of alcohol reinforces a divide between the drinking customers and the chefs, who are mostly Muslim.

The setting during ÜDTR community events, like "50 Plates of . . . ." is entirely different. The community events are free, and the exchange of knowledge, food, and experiences is reciprocal. Afternoon light floods the space, which is full of the hum and buzz of people and ingredients mixing together. The space is more open, with the tables pushed to the side, and people are gathering and moving between different kitchen islands that are hosted by chefs from different countries, including Germany. There is a frenetic energy, as people rush around to grab ingredients and chefs delegate tasks with speed, and are not shy about telling their helper participants to hurry up or adjust their chopping technique. It feels a bit like the television show Iron Chef, where chefs are in a race against time to cook a meal using whatever ingredients have been thrown at them. After all of the meals are prepared, they are spread out buffet style on the tables that line the walls. Everyone grabs a plate to mix different meals, tastes, and cultures together, and the room becomes quiet except for the sounds of gustatory pleasure, gratitude, and appreciation. Eating *with* the Other is made possible by acts of reciprocal exchange and sharing. And as Longhurst et al. (2008) note, potluck settings where bodies, ingredients, tastes, and cultures combine in unfamiliar ways can also be moments of visceral learning where eaters confront their own desires and discomforts.

In contrast to descriptions of eating *the* Other in, for example, "ethnic" restaurants that stage authenticity, through decoration, music, and dress for tourists and mainstream white diners (Heldke, 2005; Johnston & Bauman, 2015), both initiatives have created settings that encourage eating *with* the Other. In LoK, guests become family, and the power relations between eater and chef are negotiated through the norms of hospitality, gender, and domesticity. In ÜDTR, the space is designed to accommodate multiple users and identities. This seemingly neutral minimalist mid-century design also reflects the class habitus of cosmopolitan-hipster Germans and internationals in Berlin. The space is successful in creating a neutral backdrop for a variety of performances; however, depending on a diversity of factors – the actors, the arrangement of tables, the presence of alcohol, the time of day, and the mode of exchange – practices of eating *with* the Other can also slip into eating *the* Other.

Performances

The stages we have sketched earlier support and constrain a variety of culinary and social performances. These performances also change the setting and give each kitchen space a distinct vibe that can shift depending on the actors involved, the time of day, the ingredients, visceral reactions to tastes and smells, the presence of alcohol, and background sounds like music and conversation. As several scholars have noted culinary encounters across difference are often carefully staged and scripted events, designed for the pleasure and comfort of a presumed white viewer/eater (Ray, 2017; Alkon & Grosglik, 2021). In this section I seek to unpack this performance to understand the extent to which participants are able to navigate the power asymmetries of this transaction, and the tension between eating *with* the Other and eating *the* Other. I examine the creation of the performance itself, the roles different actors play in this performance, and the dramas of failed performances, discomfort, and breaking character.

LoK chefs receive training from one another and take the initiative in creating warm, welcoming environments and telling "their story." This is accomplished through individual and peer coaching, and attending LoK workshops in other women's homes to learn from others and become comfortable talking with strangers. The personal story of each chef is co-created with an anthropologist, who conducts interviews on behalf of LoK to create the booklet of recipes each participant receives. The chef and anthropologist curate each story and decide what to share. Of course, chefs can at any time go off script and share other details of their personal lives.

Creating this script and having the ability to fall back on it during culinary performances gives LoK chefs a certain amount of structure and confidence in telling their story, and protects them from unwanted questions. Being at home also allows participants to be themselves, or at least to perform those aspects of themselves that are most comfortable in this setting. In this way they have a lot of agency, but also training and support in creating performances that promote respectful culinary encounters or eating *with* the Other. The home and intimate family setting generates instant familiarity, reducing the presumed distance between self and Other that is evident in practices of exoticism.

> You meet people, you sit down with people around a table, like family and you know you turn friend, many of people they call me, they ask me or like you know we keep in touch with some people, so this is very, it's very different.
>
> (chef, League of Kitchens)

However, the familial atmosphere can also make it challenging for chefs to protect their personal boundaries – especially when being a gracious host is an important part of their personality.

> I can say, "Okay class is done", and people they enjoying, they sit always like about one hour after, an hour and a half and sometimes two hours. But now I'm like more experienced, I'm little bit more professional what I am doing. I start as woman who hosts now, I feel myself, I still am at home, you know, but as chef I'm more, you know I can deal with people.
>
> (chef, League of Kitchens)

Some guests may need a lot of attention, ignore polite signals that the workshop has ended and it's time to go, or bombard the chef with correspondence after the course. Chefs and staff exchange support and tips in confronting these kinds of challenges.

ÜDTR chefs also receive training to host workshops, from one another and the organisation. Part of the training is cooking at free community events like "50 Plates of . . .," to become practiced at interacting with a lot of different people and cooking in a more public and chaotic environment.

> You need to figure out if someone is actually able to give a cooking class and to entertain fourteen people and . . . It can be very intense. And they give a little presentation. You have to have the character for it. So that's what we try and find out.
>
> (staff, UDTR)

At paid workshop, the chefs give a PowerPoint presentation with pictures and facts about their country of origin.

> We usually have about four hours, and yes, we want to fit in all the courses we're going to cook, and we do want to fit in some talking, and we have a little presentation about Tellerrand, we have a little presentation about the cook and his or her country.
>
> (staff, ÜDTR)

The PowerPoint presentation provides a structure and frame for their narrative. However, it is more an encyclopaedia entry than a life story, and uncomfortable topics (related to war, violence, displacement, and living as a refugee) are understandably left out of the presentation.

> you can see from the people how much they want to talk. Some don't have a problem with it, some talk openly about their flight, what happened to them and so on, others don't feel like it at all and would rather concentrate on the positive things and leave the others out and I think that's where it is it's really important to respect that. Also from the guests.
>
> (staff, ÜDTR)

In the public setting of the Kitchen Hub, chefs are on display and perform their roles as teachers, experts, and chefs – providing knowledge and expertise. It may be more challenging to slip into other roles, and professionalised roles may help to preserve some distance between the chef and the paying students. This distance can get in the way of convivial encounters and developing personal connections and friendships; however, it also protects chefs who do not want to share their life story and deepest traumas with a group of strangers.

> [E]veryone knows that they're so-called refugees and where they come from and maybe how long they've been in Germany. But some people they don't want to talk about the way that they came to Germany. That's perfectly fine. No one has to know. That's a very personal thing.
>
> (staff, ÜDTR)

In some instances, paying guests are dissatisfied with the professional performance; they want more personal details, knowledge, intimacy, and connection. There are also the rare guests who came to see an authentic performance of refugee suffering. Despite the respectful setting and clearly stated mission of respectful exchange, these guests came to consume much more than food, they came to eat the Other.

> [I]f someone comes to a cooking class here and pays €75, it still doesn't mean that they can ask you everything.
>
> (staff, ÜDTR)

> Most people understand that, but there are always people in the group who go against it a bit.
>
> (staff, ÜDTR)

Their hungry questions are disruptive to the performance of culinary conviviality as "happy togetherness" and they can trigger a number of trauma responses, like shutting down or crying. I observed this response during a workshop where a white German woman, couldn't seem to stop herself from asking questions that were clearly upsetting to the chef. She only backed off after the chef began to cry. This incident led staff to reflect on what kind of support they can provide chefs, so that they don't feel alone, on exhibition, and can feel safe saying "no" to paying customers.

Although scholars have advocated for the importance of discomfort for challenging hierarchies in intercultural culinary conviviality and been critical of an overemphasis on pleasure, positive feelings, and comfort (Wise & Noble, 2016; Alkon & Grosglik, 2021), it is important to consider how this discomfort is provoked and shared and which bodies already carry more than their fair share of discomfort due to historic and ongoing traumas and injustices. For untraumatised bodies, discomfort (an unfamiliar taste, feeling out of place, etc.) can be a source of visceral learning and transformation (Longhurst et al., 2008) as well as pleasure and adventure (Heldke, 2013), while for others it can be unbearable. Critical scholars are right to critique the fact that intercultural culinary encounters are often designed for the pleasure and comfort of mainstream white bodies. However, in a convivial culinary setting comfort and discomfort are also relational, emotions become contagions traveling from body to body and create a distinct vibe that can shift from warm and friendly to cold and unsafe. In a context in which immigrants, refugees, and people of colour carry a great deal of fear and discomfort, their efforts to create "happy togetherness" and hospitality should also be appreciated as strategies for creating personal safety rather than simply reproducing power relations.

Convivial intercultural culinary encounters like "50 Plates of . . ." at ÜDTR are spaces of desire, the desire for new tastes, experiences, and friendships. But existing research on such encounters continues to centre on the desires of white mainstream culture to have "contact with the Other even as one wishes boundaries to remain intact" (hooks, 1992, p. 372). The desires of the Other are often left unexamined. Yet, these desires, which are also relational and contagious, have the potential to break down the boundaries of white mainstream culture. According to the director at the time, one of the great hopes of community events at ÜDTR is to help locals and refugees create "real friendships" and "deep personal connections" that can be mutually beneficial and can help refugees grow their personal support network.

> But such a big, big part is to have social contacts and to make new friends and to meet people who you can become like real friends with.
>
> That's sort of what we do with all our projects – to form some lasting, sustainable friendships and get people integrated properly. And it might be more important to, you know, take your mate out to a concert or go to the movies or, you know, just hang out . . . makes other stuff more, let's say, bearable.
>
> (staff, ÜDTR)

However, the hopes and desires that are mobilised at community events are actually quite challenging to satisfy in convivial encounters. One late afternoon, while preparing Syrian "pizza" together we ran out of flour. All the local shops were closed, so one of the participants volunteered to take the bus with me to buy flour at the train station shop. Leaving the buzz and hum of the Kitchen Hub, was a bit of a relief, and the cold air was refreshing. Waiting for the bus, smoking a cigarette, and standing together on the bus opened up a different setting – for performing the mundane and every day. The vibe changed from frenetic to boring, and suddenly I was the one being asked questions about my "exotic" life. My cooking partner revealed that this bus ride was actually the first time he had spent time with someone who wasn't a refugee outside of the Kitchen Hub.

Attending community events over several months, he had plenty of "encounters" with white German women, who desired contact with refugees while keeping their personal boundaries intact. But real friendship remained out of reach. Intersecting forms of social difference shape the performance and interpretation of desire at these events. The majority of the "locals" who attend these events are white, female, and German. The refugees who attend these events are almost exclusively male and Arabic. Cultural differences in gender norms and language barriers send confusing signals – and make the rules of conviviality hard to decipher. White German women wanting to create a warm and pleasant atmosphere slip into a familiar performance of flirting, laughing, and physical contact. This is confusing for the refugee men, who are performing the role of good men – and are afraid of disrespecting women or violating a cultural taboo. On the other hand, when refugee men express their desires for real friendship, this is interpreted as flirting, asking for a date, and overstepping a boundary. The normatively positive atmosphere of "happy togetherness" allows participants to avoid the "heaviness" and "negativity" of the refugee crisis, and people whose lives are in crisis. Fear of these troubles is an additional barrier to forming real friendships. Participating in ÜDTR events allows locals to "feel good" and "do good" without taking on any of the social, material, or emotional burdens of connecting with someone in crisis. The delicate balance is captured by one interviewee:

> [A]nd just to take away the heaviness a bit from this whole refugee issue. There is always negativity and sadness that is being communicated in the media. [here] It's recognizing that interesting people have come here and they also have a lot to give and we can show each other things and then cooking is just really accessible.
>
> (staff, ÜDTR)

Cooking and eating together are moments of bodily vulnerability and boundary crossing that demand implicit trust and openness in ingesting and digesting "foreign" foods, as well as the beliefs, values, and emotions attached to these foods (Hayes-Conroy & Hayes-Conroy, 2013). At LoK and ÜDTR, cooking and eating together do not simply break down boundaries and bring strangers together in "happy togetherness." Achieving understanding, connection, and friendship demands much more than good food, conviviality, commensality, and encounter. It requires the appropriate dosage of intimacy, warmth, and professionalism – for chefs and participants to safely navigate the dynamics of desire and discomfort, and the power asymmetries of this transaction. Moreover, the extent to which such practices facilitate eating *with* the Other and support personal and societal transformation, or simply reinforce existing power inequalities by eating *the* Other is still open for debate. Investigating setting and performance at LoK and ÜDTR, it became clear that there are many more factors at play that influence the overall vibe and experience of pleasure, desire, or discomfort.

Discussion: eating *the* Other or eating *with* the Other?

Whether intercultural culinary encounters are experienced as eating *with* or *of* the Other depends on the extent to which reciprocity is possible. Reciprocity is also essential to the pleasure of conviviality and commensality, where hosts and guests can share food, knowledge, and experiences – giving and receiving freely and generously. Existing power inequalities and market transactions can tip the balance of these exchanges. For example, as hooks (1992) points out – the commodification of otherness is a key element in eating the other, and erasing their difference via "consumer cannibalism that not only displaces the Other but denies the significance of that Other's history through a process of decontextualization" (1992, p. 373) by which "cultural, ethnic, and racial differences will be continually commodified and offered up as new dishes to enhance the white palate – that the Other will be eaten, consumed, and forgotten" (1992, p. 380).

Eating *the* Other is more likely to take place at paid cooking workshops. In the Kitchen Hub, the commodification of Otherness is more visible. Guests have paid a workshop fee and behave as consumers, who have paid for an experience, a delicious meal, and new knowledge about a cuisine, a person, and their culture. At paid cooking workshops in private homes, the commodification of Otherness is less visible. The commodification of culture is softened by the familial and domestic setting, and by the feeling that guests are just helping a nice auntie prepare a family meal. In addition, there are many other non-monetary exchanges that take place in the home kitchen – including the exchange of love.

> It's a very beautiful thing to do. And I feel very, very proud of what we do. And I love when I have my students and they come home and they give me all their love. Like I get it, I absorb their love. And I love when they enjoy my food. And it's not my food, it's like, we all put our energy, our love into it and I always tell them, the number one ingredient is making your food with love. Because if you don't make it like that it's not going to taste, love tastes so.
>
> (chef, LoK)

In addition, guests and hosts exchange knowledge, tips, and stories. Otherness is still on offer, but the intimacy of the encounter breaks down the voyeuristic distance that is necessary for exoticising and objectifying a person. And importantly, hosts and guests are equally interested in one another – leaving the hope of real friendship open. All of the chefs I interviewed reported making friends and growing their social and professional network through their cooking workshops. However, one important difference is that immigrant chefs in New York who are successful at making "real friends" have also had the time to establish themselves, stay with or reunite with their families, and build more settled and secure lives compared to newly arrived refugees in Berlin, who are often alone, navigating crisis, and living in shelters and unstable housing.

At ÜDTR eating *with* the Other happens most readily in free community events like "50 Plates of . . .," where locals and refugees all have something to offer – and no one is put on display to serve as culinary ambassador for their country and cuisine. This potluck setting for cooking and eating facilitates reciprocal exchanges, encounters, and the mixing of tastes, ingredients, and cultures. However, eating *with* the Other does not guarantee transformative intercultural encounters, or the possibility of "real friendship." Existing social boundaries may very well remain intact, even as strangers leave the event with greater appreciation for the food and knowledge of others. While the events are effective in staging culinary conviviality as "happy togetherness" – this vibe also smooths over the complex emotions that participants carry with them. These emotions are absorbed by the food on our plates, as sources of desire, pleasure, memory, and connection. They can be tasted, carried home in our bodies, and digested, but it would be against the social rules of "happy togetherness" for these emotions to fall off our plates. That would be sloppy eating and run the risk of creating discomfort and "trouble" for white bodies.

Conclusion

In this chapter, I have sought to explore the conditions under which convivial intercultural encounters emerge, and the extent to which such encounters contribute to the hopes of intercultural friendships and connections that can buffer the troubles of racism and xenophobia. This led me to unpack what is happening in these encounters, the setting of the stage, the possible performances and scripts, and the experiences of desire and discomfort that may result. This has revealed a number of troubles in culinary conviviality, including: the risk of eating *the* Other, provoking discomfort in traumatised bodies, and accommodating the conflicting desires of guests for both adventure, exploration, safety, and familiarity. Returning to hooks' (1992) theorisation of eating the Other as well as sociological theorisations of commensality and conviviality, I have highlighted how commodified transactions and reciprocal sharing shape these experiences and reinforce or challenge particular performances. My approach of centring the emotional experience of desire and discomfort, in both eating *the* Other and eating *with* the Other, is useful for a critical-reparative reading of community food initiatives. Following feelings in the staging and performance of culinary conviviality allows for a more nuanced analysis of the ways in which community food initiatives contain both hopes and troubles. Moreover, it offers a more relational account of the ways in which hope and trouble travel from person to person and coalesce into collective performances and vibes that support or hinder eating *with* the Other, and broader aims of personal and societal transformation through food and eating.

Acknowledgements

I would like to thank the staff and participants at the League of Kitchens and Über den Tellerrand for sharing their thoughts, experiences, and incredible meals with me. The research this chapter is based on was funded by a Horizon 2020: European Research Council Consolidator Award. Title: SHARECITY: The practice and sustainability of urban food sharing. Award No.: 646883. We are extremely grateful for this support, without which the research could not have taken place.

References

Alkon, A. H., & Grosglik, R. (2021). Eating (with) the other: Race in American food television. *Gastronomica: The Journal for Food Studies*, *21*(2), 1–13.

Davies, A. R. (2019). *Urban food sharing: Rules, tools and networks*. Bristol Policy Press.

Davies, A. R., Edwards, F., Marovelli, B., Morrow, O., Rut, M., & Weymes, M. (2017, March). Making visible: Interrogating the performance of food sharing across 100 urban areas. *Geoforum*, 86, 136–149. https://doi.org/10.1016/j.geoforum.2017.09.007

Dunbar, R. I. M. (2017). Breaking bread: The functions of social eating. *Adaptive Human Behavior and Physiology*, *3*(3), 198–211. https://doi.org/10.1007/s40750-017-0061-4

Edwards, F. (2021). Overcoming the social stigma of consuming food waste by dining at the open table. *Agriculture and Human Values*, *38*(2), 397–409.

Hayes-Conroy, A. (2010). Feeling slow food: Visceral fieldwork and empathetic research relations in the alternative food movement. *Geoforum*, *41*(5), 734–742.

Hayes-Conroy, J., & Hayes-Conroy, A. (2013). Veggies and visceralities: A political ecology of food and feeling. *Emotion, Space and Society*, 6, 81–90.

Heldke, L. (2005). But is it authentic? Culinary travel and the search for the "genuine article." *The Taste Culture Reader: Experiencing Food and Drink*, 385–394.
Heldke, L. (2013). Let's cook Thai: Recipes for colonialism. In C. Counihan & P. Van Esterik (Eds.), *Food and culture: A reader* (pp. 394–408). Routledge.
Ho, S. (2014). Craving the other: One woman's beef with cultural appropriation and cuisine. *Bitch Media*. Retrieved April 28, 2022, from www.bitchmedia.org/article/craving-the-other-0
hooks, b. (1992). *Black looks: Race and representation*. Turnaround.
Johnston, J., & Baumann, S. (2015). *Foodies*. Taylor & Francis.
Kerner, S., Chou, C., & Warmind, M. (Eds.). (2015). *Commensality: From everyday food to feast*. Bloomsbury.
Khor, S. Y. (2014). *Just eat it: A comic about food and cultural appropriation*. Retrieved April 28, 2022, from www.bitchmedia.org/post/a-comic-about-food-and-cultural-appropriation
League of Kitchens (2022). Retrieved April 28, 2022, from www.leagueofkitchens.com/
Longhurst, R., Ho, E., & Johnston, L. (2008). Using "the body" as an "instrument of research": Kimch'i and pavlova. *Area*, *40*(2), 208–217.
Marovelli, B. (2018, September). Cooking and eating together in London: Food sharing initiatives as collective spaces of encounter. *Geoforum*, 1–12. https://doi.org/10.1016/j.geoforum.2018.09.006
NYC Mayor's Office of Immigrant Affairs (2018). *State of our immigrant city*. Annual Report. https://www1.nyc.gov/assets/immigrants/downloads/pdf/moia_annual_report_2018_final.pdf
Molz, J. G. (2007). Eating difference: The cosmopolitan mobilities of culinary tourism. *Space and Culture*, *10*(1), 77–93. https://doi.org/10.1177/1206331206296383
Phull, S., Wills, W., & Dickinson, A. (2015). Is it a pleasure to eat together? Theoretical reflections on conviviality and the Mediterranean diet. *Sociology Compass*, 9(11), 977–986. https://doi.org/10.1111/soc4.12307
Ray, K. (2017). Bringing the immigrant back into the sociology of taste. *Appetite*, *119*, 41–47.
SHARECITY. Retrieved April 28, 2022, from https://sharecity.ie/
Smith, M., & Harvey, J. (2021). Social eating initiatives and the practices of commensality. *Appetite*, *161*, 105107.
Über den Tellerrand (n.d.). Retrieved April 28, 2022, from https://ueberdentellerrand.org/
Wise, A. (2012). Moving food: Gustatory commensality and disjuncture in everyday multiculturalism. *New Formations*, *74*(74), 82–107. https://doi.org/10.3898/newf.74.05.2011
Wise, A., & Noble, G. (2016). Convivialities: An orientation. *Journal of Intercultural Studies*, *37*(5), 423–431. https://doi.org/10.1080/07256868.2016.1213786

Index

Note: Page numbers in **bold** indicate a table and page numbers in *italics* indicate a figure on the corresponding page.

Printed in the United States
by Baker & Taylor Publisher Services